Nimble with Numbers

Engaging Math Experiences
to Enhance Number Sense
and Promote Practice

Grades 2 and 3

Leigh Childs, Laura Choate, and Maryann Wickett

Dale Seymour Publications
Parsippany, New Jersey

Acknowledgments

Thanks to special friends and colleagues, who helped field test these activities with their students and who gave valuable feedback and suggestions:

Eunice Hendrix-Martin

Karen Jenkins

Thanks to super kids who provided a student perspective as they tried these activities:

Children in Room 18 at Paloma Elementary School, San Marcos, California

Sarah Thompson Briggs

Megan Posner

Jenny Wickett

Senior Editor: Carol Zacny
Consulting Editor: Dorothy Murray
Production/Manufacturing Director: Janet Yearian
Production/Manufacturing Manager: Karen Edmonds
Production/Manufacturing Coordinator: Joe Conte
Design Manager: Jeff Kelly
Text design: Tani Hasegawa
Page composition: Susan Cronin-Paris
Cover design: Ray Godfrey
Art: Stefani Sadler

Dale Seymour Publications
An imprint of Pearson Learning
299 Jefferson Road; P.O. Box 480
Parsippany, NJ 07054-0480
www.pearsonlearning.com
1-800-321-3106

Copyright © 1998 by Addison Wesley Longman, Inc. All rights reserved. Printed in the United States of America. This publication is protected by Copyright and permissions should be obtained from the publisher prior to any prohibited reproduction, storage in a retrieval system, or transmission in any form or by any means, electronic, mechanical, photocopying, recording, or likewise. The publisher grants permission to individual teachers who have purchased this book to reproduce the blackline masters as needed for use with their own students. Reproduction for an entire school or school district for commercial use is prohibited. For information regarding permission(s), write to Rights and Permissions Department. This edition is published simultaneously in Canada by Pearson Education Canada.

Dale Seymour Publications® is a registered trademark of Dale Seymour, Inc.

ISBN 0-7690-2722-9

3 4 5 6 7 8 9 10-ML-05-04-03-02-01-00

Table of Contents

Introduction

Introduction to *Nimble with Numbers*	1
Planning Made Simple	2
Suggestions for Using *Nimble with Numbers*	5
Parent Support	7
Family Letter	8
Questions Sampler	9
Addition Facts Made Easy	10
Matrix of Activities	11

Addition Facts

Overview and Suggestions	13
Box Sums (S)	14
Ten Frames Addition (S)	16
Rhythm Addition (S)	17
Just the Facts 1–6 (C)	18
Seeking Sums (G)	21
Uncover (G)	23
Four Sums-in-a-Row (G)	25
Seeking Sums Practice (I)	28
Joining Neighbors (I)	30
Sum Triangles (I)	33

Subtraction Facts

Overview and Suggestions	35
Ten Frames Subtraction (S)	36
Mystery Numbers (S)	37
Number Trails (S)	39
Just the Facts 7–12 (C)	41
Making Equations (G)	44
How Many More? (G)	46
Neighbors Count (G)	49
Subtraction Squares (I)	51
Mystery Numbers Practice (I)	53
Making Equations Practice (I)	55

Money

Overview and Suggestions	57
Counting Coins (S)	58
Guess My Coins (S)	59
What's My Change? (S)	60
Sensible Cents 1–3 (C)	62
Race to 50¢ (G)	65
Pennies or Nickels? (G)	67
Who Buys? (G)	68
Buying Snacks (I)	70
Find the Combination (I)	71
What's My Change? (I)	72

Place Value

Overview and Suggestions	73
100 Chart Pieces (S)	74
Arrow Math (S)	76
Place Value Paths (S)	77
Where Will It Fit? (S)	79
What's in That Place? 1–6 (C)	81
100 Chart Cover (G)	84
100 Chart Paths (G)	85
Will It Fit? (G)	86
Where? (G)	88
What Numbers Are Missing? (I)	90
Staircase Number Puzzles (I)	93
Creating Numbers (I)	95

Addition

Overview and Suggestions	97
Tossing Sums (S)	98
Add It Up (S)	100
Sum It Up 1–6 (C)	102
18 Plus (G)	105
26 Plus (G)	107
Slide and Sum (G)	108
Target 50 (G)	110
Loop Addition (I)	112
Addition Trees (I)	114
Making Sums (I)	116
Searching for 30 (I)	118

Subtraction

Overview and Suggestions	119
Three Ten Frames (S)	120
Make the Difference (S)	123
Finding Differences (S)	126
What's the Difference? 1–6 (C)	127
Subtract and Travel (G)	130
Differences Count (G)	133
Difference Bingo (G)	136
Subtraction Squares (I)	139
Finding Pairs (I)	141
Rearrange and Find (I)	143

Blackline Masters

100 Chart	146
Coins	147
Digit Cards	148
Digit Squares	149
Dot Patterns	150
Number Cubes (1–6, 3–8)	151
Number Cubes (4–9, blank)	152
Spinners (1–6, blank)	153
Subtract and Travel Form	154
Subtraction Squares Form	155
Ten Frames	156
Addition Chart	157

Answer Key 158

Introduction to *Nimble with Numbers*

Why This Book?

National recommendations for a meaning-centered, problem-solving mathematics curriculum place new demands on teachers, students, and parents. Students need a facility with numbers and operations to achieve success in today's mathematics programs. Basics for students today require a broadening of the curriculum to include all areas of mathematics. Students are being asked to demonstrate proficiency not just in skills, but in problem solving, critical thinking, conceptual understanding, and performance tasks. Consequently, the reduced time teachers devote to number must be thoughtful, selective, and efficient.

This book fulfills the need for high-quality, engaging math experiences that provide meaningful practice and further the development of number sense and operation sense. **These activities are designed to help students practice number concepts previously taught for understanding in a variety of contexts.** Besides meeting the need for effective practice, *Nimble with Numbers*:

- provides a variety of adaptable formats for essential practice;
- supplements and enhances homework assignments;
- encourages parent involvement in improving their child's proficiency with basic facts and computation; and
- provides motivating and meaningful lessons for a substitute teacher.

Criteria for Preferred Activities

For efficient use of time devoted to the number strand, the book focuses on activities that are:

- Inviting (encourages participation)
- Engaging (maintains interest)
- Simple to learn
- Repeatable (able to reuse often and sustain interest)
- Open-ended, allowing multiple approaches and solutions
- Easy to prepare
- Easy to adapt for various levels
- Easy to vary for extended use

In addition, these activities:

- Require a problem-solving approach
- Improve basic skills
- Enhance number sense and operation sense
- Encourage strategic thinking
- Promote mathematical communication
- Promote positive attitudes toward mathematics as mathematical abilities improve

Planning Made Simple

Organization of Book

The activities of this book are divided into eight sections which cover high-priority number topics for second graders and third graders. The first two sections review the addition and subtraction facts. Procedures to minimize the number of addition facts to learn are outlined at the end of the introductory section, on page 10. Share these strategies with parents to help them use their available time more efficiently.

The third section reinforces money concepts which give students an opportunity to practice real-life skills. The money section addresses a high-interest topic and provides a relevant lead-in to the place value section. By devoting more time to the first four sections, students should experience more success with the addition and subtraction computation sections. Throughout all sections, we make an obvious attempt to promote mental computation.

Each section begins with an overview and suggestions to highlight the activities and provide some time-saving advice. The interactive activities identify the specific topic practiced (Topic), the objective (Object), the preferred grouping of participants (Groups), and the materials required (Materials). At the end of many activities, "Making Connections" questions promote reflection and help students make mathematical connections. Tips are often included to provide helpful implementation suggestions and variations. Needed blackline masters are included with the activity or in the Blackline Masters section at the end of the book.

The introductory section concludes with a Matrix of Activities. The repeatable Sponges and Games are listed alphabetically with corresponding information to facilitate their use.

Types of Activities

The book contains activities for whole group, small groups, pairs, and individuals. Each section provides:

- Sponges (S)
- Skill Checks (C)
- Games (G)
- Independent Activities (I)

Sponges

Sponges are enriching activities for soaking up spare moments. Use Sponges with the whole class or with small groups as warm-up activities, or during spare time to provide additional math practice. Sponges usually require little or no preparation and are short in duration (3–15 minutes). These appealing Sponges are repeatable and, once they become familiar, can be student-led. Students are motivated to finish a task when they know a favorite Sponge will follow.

Skill Checks

The Skill Checks in each section provide a way to show students' improvement to the parents as well as to the students. With the exception of the *Sensible Cents* sheets, each page is designed to be duplicated and cut in half, providing six comparative records for each student. Before answering the ten problems in each Skill Check, students should respond to the starter task following the STOP sign. These starter tasks are intended to promote mental computation and build number sense. Some teachers believe their students perform better on the Skill Checks if the responses to the STOP task are shared and discussed before students solve the remaining ten problems. Most students will complete a Skill Check in 10 to 15 minutes. The concluding extension problem, labeled "GO ON," accommodates those students who finish early. We recommend that early finishers be encouraged to create similar problems for others to solve. By having students share and discuss their approaches and responses to the STOP task and to some of the problems, teachers help students discover more efficient mental computation strategies.

Games

Initially a new Game might be modeled with the entire class, even though Games are intended to be played by pair players or small groups after the rules are understood. ("Pair players" refers to players who collaborate to play against another pair. This recommended arrangement promotes mathematical thinking and communication as students collaborate to develop and share successful strategies.) An excellent option is to share the Game with a few students who then teach the Game to others. Some Games in this book are intended for collaboration only and are not competitive. To facilitate getting started, teachers may recommend some procedure for identifying the first pair or player. Many Games include easy versions as well as more challenging versions, especially in the Subtraction section. Most Games require approximately 20 to 45 minutes of playing time. Games are ideal for home use since they provide students with additional practice and reassure parents that the number strand continues to be valued. When sending gameboards home, be sure to include the directions.

Games and Sponges

Games and Sponges provide students with a powerful vehicle for assessing their own mathematical abilities. During the Games, students receive immediate feedback that allows them to revise and to correct inefficient and inadequate practices. Because of their appealing and repeatable nature, these Sponges and Games are valuable as center activities. Sponges and Games differ from the Independent Activities since they usually need to be introduced by a teacher or leader.

Independent Activities

Independent Activity sheets provide facts and computation practice for students. These sheets are designed to encourage practice of many more facts than would seem apparent at first glance. Some Independent Activity sheets allow multiple solutions. Most students will complete an Independent Activity sheet in 15 to 30 minutes. Independent Activity sheets can be completed in class or sent home as homework. Many Independent Activities provide two versions to accomodate different levels of difficulty and can be easily modified to provide additional practice.

Suggestions for Using *Nimble with Numbers*

Materials Tips

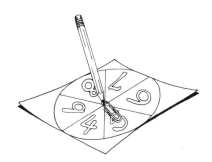

An effort has been made to minimize the materials needed. When appropriate, blackline masters are provided. The last section of the book contains more generic types of blackline masters, including patterns for various number cubes. The six-sectioned **spinners** (p. 153) can substitute for a number cube or die. The blank spinner can be used for the specially marked number cubes (3–8 or 3–3–4–4–5–5). A simple spinner, like the one shown, can be assembled using one of the blackline master spinner bases, a paper clip, and a pencil. To reduce the noise and confine the area where cubes are rolled, use a box with felt glued to its bottom or lid.

Some activities use the Digit Squares (p. 149). The familiar sets of 0–9 number tiles substitute well for Digit Squares. If not available, take time now to duplicate a Digit Square set on card stock for each child.

A few activities require Digit Cards (p. 148). Digit Card sets should also be duplicated on card stock. Digit Cards are also needed for some class sponges. Teachers should cut two sets of Digit Cards apart, place them in an appropriate container (paper sack, coffee can, or margarine tub), and store in a handy place. Some children may need Dot Patterns (p. 150) substituted for Digit Cards. Children will have more success with the money activities if they have access to play coins or coin representations (p. 147).

Various materials work as markers on gameboards—different types of beans, multicolored cubes, buttons, counters, or transparent bingo chips (our preference due to the see-through feature). For some activities children will need scratch paper and pencils. It is assumed that an overhead projector is available, but a chalkboard may be substituted.

Recommended Uses

The repeatable nature of these activities makes them ideal for continued use at home. Encouraging children to use these activities at home serves a dual purpose: parents are able to assist their children in gaining competence with the facts and with mental computation, and parents are reassured as they see the familiar basics practiced. To support your work in this area, we have included a parent letter and a list of helpful open-ended questions.

Besides being a source for more familiar homework, these activities offer a wide variety of classroom uses. The activities can be effectively used by substitute teachers as rainy-day options or for a change of pace. Many activities are short-term and require little or no preparation, making them ideal for soaking up spare moments at the end or beginning of a class period. They also work well as choices for center or menu activities. When students are absent from school, include these activities in independent work packets. You may package these activities in manila envelopes or self-closing transparent bags to facilitate frequent and easy checkout. To modify the activities and to accommodate the needs of your students, you may easily change the numbers, operations, and directions.

Getting the Most from These Activities

It is important to focus on increasing childrens' awareness of the mathematics being learned. To do this, pose open-ended questions that promote reflection, communication, and mathematical connections. For example, after using *How Many More?*, one colleague asked her students, "What mathematics are you doing?" Her third graders identified subtraction facts, addition facts, finding parts, and algebra ("finding an unknown"). After using *Neighbors Count*, a colleague asked her students to estimate the total number of facts they had practiced. The range of responses was great.

Having students work together as pair players is of great value in increasing student confidence. While working this way, students have more opportunities to communicate strategies and to verbalize thinking. When asked to identify and to share their successful game strategies verbally and in writing, students grow mathematically. Also, it is worthwhile to ask students to improve these activities or to create different versions of high-interest games.

Good questions help children make sense of mathematics, build their confidence, and encourage mathematical thinking and communication. A list of helpful, sample questions appears on page 9. Since the teacher's or parent's response impacts learning, we have included suggestions for responding. Share this list with parents for their use as they assist students with these activities and other unfamiliar homework tasks. This list was created by Leigh Childs for parent workshops and for inclusion in the California Mathematics Council's *They're Counting on Us, A Parent's Guide to Mathematics Education*. We have adapted the list for use with this book.

Parent Support

Parent Involvement

Since most parents place a high priority on attention to the number strand, they will appreciate the inviting and repeatable activities in this book. Because most parents willingly share the responsibility for repeated, short periods of practice, the following items are designed to promote parent involvement: *Family Letter* (p. 8), *Questions Sampler* (p. 9), and *Addition Facts Made Easy* (p. 10). The first home packet might include the *Family Letter,* the *Questions Sampler,* and *Seeking Sums* (pp. 21–22). To facilitate use, students should take home one set of Digit Cards (p. 148). Since many students will benefit by practicing the addition and subtraction facts, the next home packet might include *Addition Facts Made Easy* (p. 10), the *Addition Chart* (p. 157), and *Uncover* (pp. 23–24) or *Making Equations* (pp. 44–45). This packet should include materials for making two Number Cubes (p. 151 for *Uncover*) or Digit Squares (p. 149 for *Making Equations*). Advise students and their families to keep the Number Cubes and Digit Squares in a safe place for frequent use throughout the school year.

Students enjoy and benefit from repeated use of *Number Trails* (pp. 39–40) and *Guess My Coins* (p. 59) . These Sponges lend themselves to home packets as well. The advantage of Sponges, unlike Games, is that many can be experienced while a monitoring family member prepares dinner, packs lunches, or attends to other household tasks.

Concluding Thought

We hope that by using these materials, your students will develop more positive feelings toward mathematics as they improve their confidence and number competence.

Family Letter

Dear Family,

To be prepared to work in the 21st century, all students need to be confident and competent in mathematics. Today the working world requires understanding of all areas of mathematics including statistics, logic, geometry, and probability. To be successful in these areas, students must know their basic facts and be able to compute. It is important that we be more efficient and effective in the time we devote to arithmetic. You can help your child in this area.

Throughout the school year, our mathematics program will focus on enhancing your child's understanding of number concepts. However, students must devote time at school and at home to practice and to improve these skills. Periodically, I will send home activities and related worksheets that will build number sense and provide much needed practice. These games and activities have been carefully selected to engage your child in practicing more math facts than usually answered on a typical page of drill or during a flash-card session.

By using the enclosed *Questions Sampler* during homework sessions, you will be able to assist your child without revealing the answers. The questions are categorized to help you select the most appropriate questions for your situation. If your child is having difficulty getting started with a homework assignment, try one of the questions in the first section. If your child gets stuck while completing a task, ask one of the questions from the second section. Try asking one of the questions from the third section to have your child clarify his or her mathematical thinking.

Good questions will help your child make sense of the mathematics, build confidence, and improve mathematical thinking and communication. I recommend posting the *Questions Sampler* in a convenient place, so that you can refer to it often while helping your child with homework.

Your participation in this crucial area is most welcome.

Sincerely,

Questions Sampler

Getting Started

How might you begin?

What do you know now?

What do you need to find out?

While Working

How can you organize your information?

How can you make a drawing (model) to explain your thinking?

What approach (strategy) are you developing to solve this?

What other possibilities are there?

What would happen if . . . ?

What do you need to do next?

What assumptions are you making?

What patterns do you see? . . . What relationships?

What prediction can you make?

Why did you . . . ?

Checking Your Solutions

How did you arrive at your answer?

Why do you think your solution is reasonable?

What did you try that didn't work?

How can you convince me your solution makes sense?

Expanding the Response

(To help clarify your child's thinking, avoid stopping when you hear the "right" answer and avoid correcting the "wrong" answer. Instead, respond with one of the following.)

Why do you think that?

Tell me more.

In what other way might you do that? What other possibilities are there?

How can you convince me?

Addition Facts Made Easy

Since the tables go from 1 to 10, it appears there are 100 "facts" to memorize. However . . .

If you eliminate the easy 1s and 10s, you **eliminate 36 facts.**

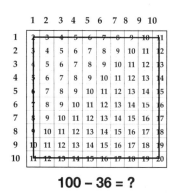

100 − 36 = ?

If you eliminate the 2s which also seem easy, you **eliminate another 15 facts.**

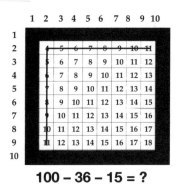

100 − 36 − 15 = ?

The **doubles,** like 4 + 4 and 6 + 6, seem easy to remember, so you can **eliminate another 7.**

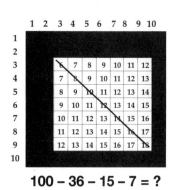

100 − 36 − 15 − 7 = ?

When you know the **doubles plus one** facts, you can **eliminate another 12.**

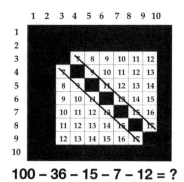

100 − 36 − 15 − 7 − 12 = ?

After you have learned the 9s, you can *reduce* the number of facts **by another 10.**

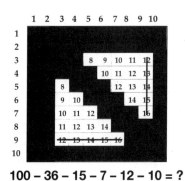

100 − 36 − 15 − 7 − 12 − 10 = ?

We still have duplicates like "3 + 5" and "5 + 3." So, we can cut the remaining 20 facts in half, **eliminating 10 more.**

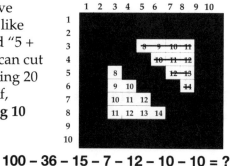

100 − 36 − 15 − 7 − 12 − 10 − 10 = ?

These remaining **"difficult" facts** are left to memorize.

| 5 + 3 | 6 + 3 | 7 + 3 | 8 + 3 | 6 + 4 |
| 7 + 4 | 8 + 4 | 7 + 5 | 8 + 5 | 8 + 6 |

It is **highly recommended** that students practice the **related subtraction facts** for quick recall. (For example, 8 − 3 and 8 − 5 are the related subtraction facts for 5 + 3.) For the ten facts listed above, there are 20 related subtraction facts.

Matrix of Games and Sponges

G = Games   S = Sponges

Type	Title	Topic	Page	Materials	Class	Groups	Pairs
G	100 Chart Cover	Place Value	84	Digit Cards, Markers, Form p. 146		✓	✓
G	100 Chart Paths	Place Value and Subtraction	85	Digit Cards, Markers, Form p. 146			✓
S	100 Chart Pieces	Number Relationships	74	Markers, Forms pp. 75 & 146	✓		
G	18 Plus/26 Plus	Addition	105	Number Cubes, Markers, Gameboard p. 106-107			✓
S	Add It Up	Mental Addition	100	Digit cards including Transparent Set, Form p. 101	✓	✓	
S	Arrow Math	Number Sense to 100	76	Markers, Form p. 146	✓	✓	
S	Box Sums	Addition Facts	14	Digit Cards including Transparent Set, Form p. 15	✓	✓	
S	Counting Coins	Mental Addition of Coins	58	Overhead Coins	✓	✓	
G	Difference Bingo	Subtraction	136	Digit Cards, Markers, Gameboards pp. 137-138		✓	✓
G	Differences Count	Subtraction	133	Digit Cards, Forms pp. 134-135		✓	✓
S	Finding Differences	Mental Subtraction	126	Chalkboard or Overhead Projector	✓		✓
G	Four Sums-in-a-Row	Addition Facts	25	Paper Clips, Markers, Gameboards pp. 26-27		✓	✓
S	Guess My Coins	Mental Addition of Coins	59	Prepared Containers, Play Coins	✓	✓	
G	How Many More?	Subtraction Facts	46	Number Cube, Markers, Gameboards pp. 47-48		✓	✓
S	Make the Difference	Mental Subtraction	123	Digit Cards, Forms pp. 124-125		✓	✓
G	Making Equations	Subtraction and Addition Facts	44	Digit Squares, Form p. 45	✓		✓
S	Mystery Numbers	Subtraction and Addition Facts	37	Digit Squares, Self-stick Shapes, Form p. 38		✓	✓
G	Neighbors Count	Subtraction and Addition Facts	49	Number Cubes, Markers, Gameboard p. 50		✓	✓
S	Number Trails	Subtraction and Addition Facts	39	Form p. 40	✓	✓	
G	Pennies or Nickels?	Addition of Nickels and Pennies	67	Number Cubes, Play Coins		✓	✓
S	Place Value Paths	Number Sense	77	Digit Cards, Forms pp. 78 and 100	✓		✓
G	Race to 50¢	Exchanging and Adding Coins	65	Number Cubes, Markers, Play Coins, Gameboard p. 66		✓	✓
S	Rhythm Addition	Addition Facts	17		✓	✓	
G	Seeking Sums	Addition Facts	21	Digit Cards, Markers, Gameboard p. 22		✓	✓
G	Slide and Sum	Mental Addition to 50	108	Markers, Gameboard p. 109			✓
G	Subtract and Travel	Subtraction	130	Number Cubes, Markers, Gameboards pp. 131-132, 154			✓
G	Target 50	Mental Addition	110	Digit Squares, Form p. 111			✓
S	Ten Frames Addition	Addition Facts	16	Markers, Form p. 156		✓	

Introduction 11

Matrix of Games and Sponges (cont.)

G = Games S = Sponges

Type	Title	Page	Materials	Topic	Class	Groups	Pairs
S	Ten Frames Subtraction	36	Markers, Form p. 156	Subtraction Facts	✓		
S	Three Ten Frames	120	Form pp. 121-122	Subtraction	✓	✓	
S	Tossing Sums	98	Form p. 99	Mental Addition	✓	✓	
G	Uncover	23	Number Cubes, Markers, Gameboard p. 24	Addition Facts			✓
S	What's My Change?	60	Overhead Coins, Play Coins, Form p. 61	Adding and Subtracting Coins	✓	✓	
S	Where Will It Fit?	79	Number Cubes, Form p. 80	Place Value	✓		✓
G	Where?	88	Number Cubes, Form p. 89	Place Value			✓
G	Who Buys?	68	Markers, Play Coins, Gameboard p. 69	Mental Addition of Coins			✓
G	Will It Fit?	86	Digit Cards, Form p. 87	Place Value			✓

Nimble with Numbers

Copyright© Dale Seymour Publications®

Addition Facts

Assumptions The addition facts have previously been taught and reviewed with an emphasis on understanding. Concrete objects and visual models, such as Ten Frames and dominoes, have been used extensively. Students have received assistance in categorizing the facts, so that they will learn the few facts not quickly recalled. (See *Addition Facts Made Easy*, p. 10.)

Section Overview and Suggestions

Sponges

Box Sums pp. 14–15

Ten Frames Addition p. 16

Rhythm Addition p. 17

These open-ended warm-ups actively engage children in practicing many addition facts. *Ten Frames Addition* reinforces a valuable visual model. The frequent use of these Sponges will ensure greater success with the Games and Independent Activities in this section.

Skill Checks

Just the Facts 1–6 pp. 18–20

These provide a way to see children's improvement with the addition facts. Copies may be cut in half so that each Check may be used at a different time. Remember to have children respond to STOP, the number sense task, before they solve the ten problems.

Games

Seeking Sums pp. 21–22

Uncover pp. 23–24

Four Sums-in-a-Row pp. 25–27

These open-ended repeatable Games actively involve children in practicing many addition facts. In *Seeking Sums*, pairs collaborate to complete the activity. Challenging *Uncover* and *Four Sums-in-a-Row* encourage flexible thinking. The second *Four Sums-in-a-Row* gameboard allows children to practice more difficult facts. All Games promote mental computation as children enhance their strategic thinking skills.

Independent Activities

Seeking Sums Practice pp. 28–29

Joining Neighbors pp. 30–32

Sum Triangles pp. 33–34

Each Independent Activity has two or three worksheets which allow children to progress to more difficult addition facts. Success on *Seeking Sums Practice* will be enhanced by repeated experiences with the corresponding Game. *Seeking Sums Practice* and *Joining Neighbors* provide initial practice with the helpful, visual dot patterns format. Activities may have more than one solution due to their open-ended nature.

Box Sums

Topic: Addition Facts

Object: Mentally find sums in designated sections of a box square.

Groups: Whole class or small group

Materials

- transparency of *Box Sums* activity sheet, p. 15
- transparency of Digit Cards, p. 148
- 5-by-8-inch index card
- set of Digit Cards for each child, p. 148

Tip If children need to gain more confidence adding digits, replace Digit Cards with Dot Pattern Cards.

Directions

1. The leader displays *Box Sums* activity sheet with Digit Cards placed in each of the four sections.
2. Using an index card, the leader hides the right half of the activity sheet and asks children, "What's the sum? What did you do mentally to find the sum?"
3. The leader continues this process for each half section. Children respond by showing the sums using their Digit Cards.
4. The leader displays entire activity sheet and asks, "How many altogether? What did you do mentally to find the sum?"
5. The leader places new combinations of Digit Cards on the Box Sums activity sheet and displays different half sections. Children show the sum with their Digit Cards.
6. These steps can be repeated with a variety of combinations.

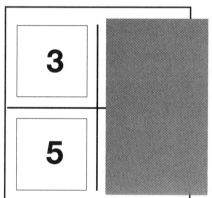

Making Connections

Promote reflection and make mathematical connections by asking:

- What strategies did you use to quickly identify sums?

Box Sums

Sponge   Addition Facts

Ten Frames Addition

Topic: Addition Facts

Object: Visualize and verbalize sums.

Groups: Whole class or small group

Materials

- transparency of Ten Frames, p. 156
- counters (cubes are easily handled)

Tip For less-confident children, begin by combining different amounts that total ten.

Directions

1. The leader displays eight counters in the upper Ten Frame and four counters in the lower Ten Frame.
2. The leader asks children to identify the number of counters in the top Ten Frame, and in the lower Ten Frame. (8, 4)
3. The leader asks children, "How many counters altogether?" (12) "How did you get this total?"
4. The leader changes only the number of counters in the lower Ten Frame and asks a similar series of questions. The leader continues changing the number of counters in the lower Ten Frame and asking questions.
5. The leader changes the top Ten Frame to nine counters (or some new amount) and repeats the sequence of questions.

Making Connections

Promote reflection and make mathematical connections by asking:

- How do Ten Frames help you identify the correct sums?

Rhythm Addition

Topic: Addition Facts

Object: Name sums while maintaining the rhythm.

Groups: Whole class or small group

Tip Set a slow pace to allow for greater success.

Directions

1. The leader identifies a pathway around the classroom that includes all children.
2. The entire class practices a "2 slaps, 2 claps, 4 snaps" continuing rhythm.
3. When everyone seems to have the rhythmic pattern, the leader begins by stating an addition fact during the "snapping stage." Without interrupting the rhythmic activity, the first child responds with the sum timed with the following set of snaps. (All responses and naming of facts occur during the "snapping stage.")
4. The leader continues this warm-up with each child answering a stated fact. The goal is to go all around the classroom or group without breaking the rhythmic pattern and by responding in the expected timely manner.

Making Connections

Promote reflection and make mathematical connections by asking:

- What suggestions could be made to ensure success for the entire class?

Sponge

Date _____ Name _____

Just the Facts 1

 Don't start yet! Star two problems that may have odd answers.

1. 3
  + 4
  ___

2. 6
  + 3
  ___

3. 7
  + 5
  ___

4. 8
  + 8
  ___

5. 2 + 5 = _____

6. 6 + 1 = _____

7. 8 + 3 = _____

8. 4 + 9 = _____

9. 7 + 9 = _____

10. 6 + 8 = _____

Go On ➡ What numbers come next? 4, 8, 12, _____ , _____

Date _____ Name _____

Just the Facts 2

 Don't start yet! Star the problem in Row 1 that may have the largest answer.

1. 2
  + 5
  ___

2. 5
  + 5
  ___

3. 8
  + 4
  ___

4. 9
  + 8
  ___

5. 6 + 2 = _____

6. 4 + 5 = _____

7. 7 + 4 = _____

8. 8 + 5 = _____

9. 8 + 8 = _____

10. 9 + 7 = _____

Go On ➡ Write three facts that equal 11.

Nimble with Numbers **Skill Checks**

Date _____ Name _____

Just the Facts 3

 Don't start yet! Star two problems that may have even answers.

1. 3
   + 5
   ___

2. 7
   + 3
   ___

3. 6
   + 5
   ___

4. 7
   + 8
   ___

5. 4 + 4 = _____

6. 7 + 2 = _____

7. 6 + 4 = _____

8. 9 + 3 = _____

9. 6 + 7 = _____

10. 9 + 9 = _____

 What numbers come next? 5, 10, 15, _____ , _____ , _____

Date _____ Name _____

Just the Facts 4

 Don't start yet! Star two problems that may have answers less than 10.

1. 4
   + 2
   ___

2. 2
   + 8
   ___

3. 6
   + 6
   ___

4. 8
   + 6
   ___

5. 3 + 3 = _____

6. 2 + 7 = _____

7. 5 + 7 = _____

8. 9 + 4 = _____

9. 6 + 9 = _____

10. 8 + 7 = _____

Go On Write three facts that equal 12.

Skill Checks Addition Facts

Date _____ Name _____

Just the Facts 5

STOP Don't start yet! Star the problem that may have the smallest answer.

1. 2
   + 2
   ___

2. 3
   + 6
   ___

3. 9
   + 3
   ___

4. 5
   + 9
   ___

5. 7 + 2 = _____

6. 1 + 8 = _____

7. 6 + 5 = _____

8. 5 + 8 = _____

9. 7 + 7 = _____

10. 8 + 9 = _____

Go On What numbers come next? 3, 6, 9, _____, _____, _____

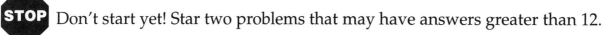

Date _____ Name _____

Just the Facts 6

STOP Don't start yet! Star two problems that may have answers greater than 12.

1. 5
   + 1
   ___

2. 4
   + 6
   ___

3. 9
   + 2
   ___

4. 4
   + 9
   ___

5. 3 + 4 = _____

6. 3 + 7 = _____

7. 7 + 5 = _____

8. 4 + 8 = _____

9. 9 + 8 = _____

10. 8 + 7 = _____

Go On Write three facts that equal 13.

Nimble with Numbers — Skill Checks

Seeking Sums

Topic: Addition Facts

Object: Equal the target sums.

Groups: Pairs

Materials for each group

- Digit Cards 1–6, p. 148
- *Seeking Sums* gameboard, p. 22
- 18 markers

Tip For children ready for a greater challenge, use digits 1 through 9.

Directions

1. A pair draws four Digit Cards and displays them at the top of their *Seeking Sums* gameboard.

2. The two children collaborate while taking turns using any of the four displayed numbers to make the sums 1 through 18. When a child identifies a way to show a sum, he or she covers that sum with a marker.

Example: If 2, 3, 5, and 6 are displayed, 5 could be made two different ways (combining 2 and 3 or with the 5 card alone).

3. The pair continues until agreeing that no other sums can be made.

4. If time allows, clear the gameboard, mix the six Digit Cards, and begin another round.

Making Connections

Promote reflection and make mathematical connections by asking:

- Which sums were not possible? Why?
- Which sums could be made more than one way?

1	2	3	4	⑤	6
7	8	9	10	11	12
13	14	15	16	17	18

Displayed digit cards: 2, 3, 5, 6

Seeking Sums

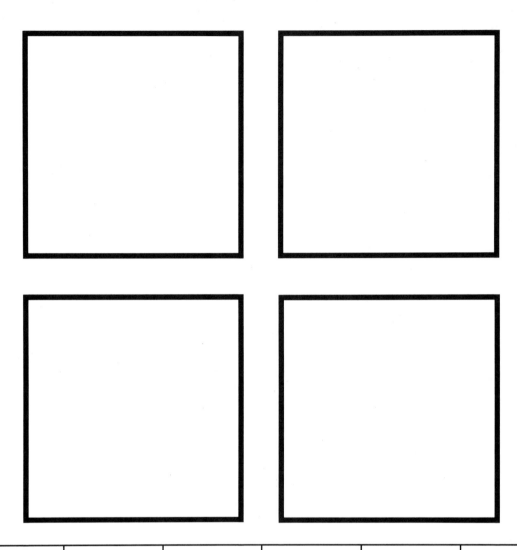

Uncover

Topic: Addition Facts

Object: Uncover the most numbers.

Groups: Pair players or 2 players

Materials for each group
- *Uncover* gameboard, p. 24
- 2 Number Cubes (1–6), p. 151
- 24 transparent markers

Tip As children gain confidence, have them determine their scores by totaling the numerical values of the uncovered numbers. Winners have lower scores.

Directions

1. Using markers, pairs cover every numbered space along their number strips on the gameboard.
2. The first pair rolls the number cubes and adds the numbers on the cubes. The pair can uncover the sum or any combination of addends that equals the rolled sum.

   Example: If 2 and 4 are rolled, the pair can uncover 6 or any combination that makes 6 (1 + 5, 1 + 2 + 3, or 2 + 4).

3. Pairs alternate turns, rolling number cubes and uncovering sums or addends on their number strips.
4. When a pair can no longer use the covered numbers to make a sum or combination, play stops for that pair. When neither pair can uncover any more numbers, the game ends.
5. The pair who uncovers the most numbers wins.

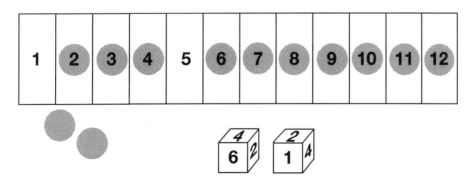

Making Connections
Promote reflection and make mathematical connections by asking:
- Which sums did you prefer to roll? Why?
- What strategies helped you uncover more markers?

Uncover

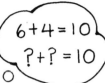

Four Sums-in-a-Row

Topic: Addition Facts

Object: Cover four numbers in a row.

Groups: 2 pair players

Materials for each group

- *Four Sums-in-a-Row A* gameboard, p. 26
- 2 paper clips
- different markers for each pair

Tips For players feeling insecure with the facts, allow three in a row to win. Use Gameboard B to practice sums through 18.

Directions

1. The first pair places two paper clips at the bottom of the gameboard, indicating two addends, adds the two numbers, and places a marker on the resulting sum.
2. The other pair moves *only one* of the paper clips to a new addend, adds the two indicated numbers, and places a marker on that sum. (It is permissible to have two paper clips on the same addend.)
3. Play continues with pairs alternating turns, moving one paper clip each time, adding the numbers, stating the fact, and placing markers on the gameboard.
4. The first pair to have four markers in a row horizontally, vertically, or diagonally wins.

Making Connections

Promote reflection and make mathematical connections by asking:

- Was it difficult to block your opponent? Why or why not?
- What strategies helped you line up your markers in a row?

10	8	7	4	11
11	9	12	9	7
6	11	8	10	3
8	5	7	11	12
5	10	12	9	6

0 1 2 3 4 5 6 7 8 9

Four Sums-in-a-Row A

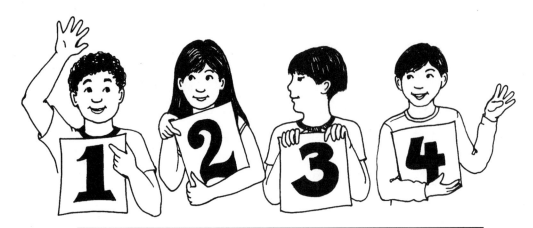

10	8	7	4	11
11	9	12	9	7
6	11	8	10	3
8	5	7	9	12
5	10	12	10	6

0 1 2 3 4 5 6 7 8 9

Four Sums-in-a-Row B

11	6	14	15	13
13	17	10	18	16
8	13	12	14	17
10	15	16	7	12
14	12	9	11	18

0 1 2 3 4 5 6 7 8 9

Seeking Sums Practice I

1 _____   _____   _____

Which of the sums below can be shown using the dot patterns above?
If you find a way, record your solution.

3 _____   6 _____   9 _____

4 _____   7 _____   10 _____

5 _____   8 _____   11 _____

---

1 _____   _____   _____

Which of the sums below can be shown using the dot patterns above?
If you find a way, record your solution.

4 _____   9 _____

6 _____   10 _____

7 _____   11 _____

8 _____   12 _____

Nimble with Numbers — **Independent Activity**

Date _____   Name _____

Seeking Sums Practice II

    [dot pattern: 4] [dot pattern: 6]

| 1   _____   _____   _____

Which of the sums below can be shown using the dot patterns above?
If you find a way, record your solution.

3 _____   7 _____   10 _____

5 _____   8 _____   11 _____

6 _____   9 _____   12 _____

| 1 | | 2 | | 5 | | 6 |

Which of the sums below can be shown using the numbers above?
If you find a way, record your solution.

6 _____   10 _____

7 _____   11 _____

8 _____   12 _____

9 _____   13 _____

Independent Activity   Addition Facts

Joining Neighbors I

Use the dot patterns that are next to each other to make the sums shown.
The first one is done for you.

6 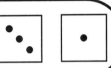

9

7

11

8

10

12

9

11

12

Nimble with Numbers   **Independent Activity**

Joining Neighbors II

Use the numbers that are next to each other to make the sums shown.

8	4	6	3	2	3
11	4	2	5	3	4
13	5	3	6	7	4
10	3	4	2	4	5
12	4	2	5	3	2

12	6	5	4	3	6	1
10	2	3	4	2	1	4
13	4	2	3	4	3	5
12	5	3	2	4	1	5
13	3	4	4	3	2	5

Independent Activity

Addition Facts

Date _____   Name _____

Joining Numbers III

Use the numbers that are next to each other to make the sums shown.

(12) 5 4 6 3 3   (14) 4 6 3 5 4

(15) 6 3 4 8 5   (13) 6 3 5 1 4

(11) 7 2 5 2 4   (15) 5 3 4 3 6

(13) 5 3 5 4 6   (18) 7 5 4 7 7

(16) 6 4 3 5 4 3

(17) 7 3 4 5 2 3

(17) 4 6 3 4 2 5

(18) 7 4 5 6 3 6

32 Nimble with Numbers   Independent Activity

Date _____   Name _____

Sum Triangles I

Make the sum of the numbers on each side equal the number in the center of the triangle.

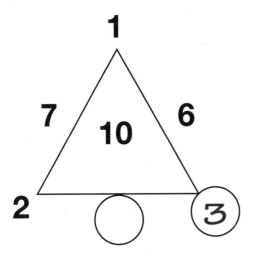
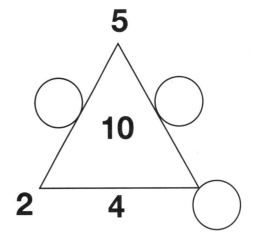
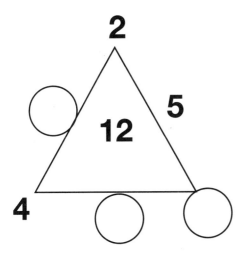
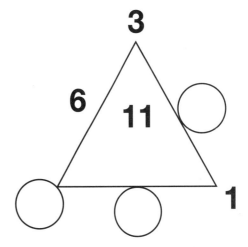
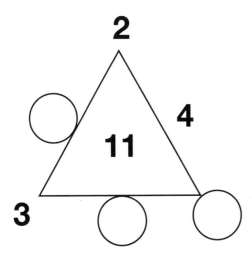
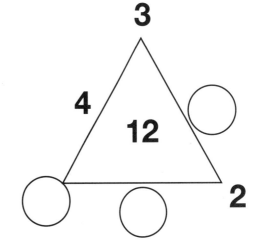

Independent Activity   Addition Facts   33

Date _____   Name _____

Sum Triangles II

Make the sum of the numbers on each side equal the number in the center of the triangle.

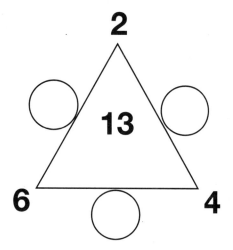
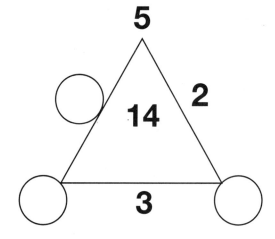
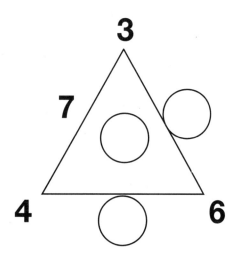
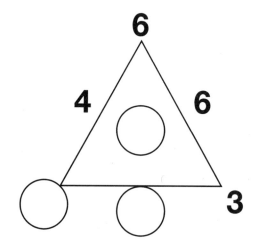
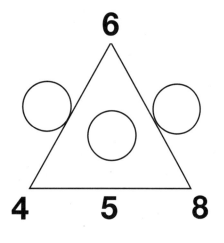
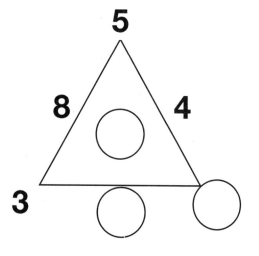

34 Nimble with Numbers   Independent Activity

Subtraction Facts

Assumptions The subtraction facts have previously been taught and reviewed with an emphasis on understanding. Concrete objects and visual models, such as counters and Ten Frames, have been used extensively. Students have received assistance in categorizing the facts, so that they will learn the few facts not quickly recalled. (*See Addition Facts Made Easy*, p. 10.)

Section Overview and Suggestions

Sponges

Ten Frames Subtraction p. 36

Mystery Numbers pp. 37–38

Number Trails pp. 39–40

These open-ended warm-ups actively engage children in practicing many subtraction and addition facts. The frequent use of these Sponges will ensure greater success with the Games and Independent Activities in this section.

Skill Checks

Just the Facts 7–12 pp. 41–43

These provide a way to help parents, children, and you see children's improvement with the subtraction facts. Copies may be cut in half so that each Check may be used at a different time. Remember to have children respond to STOP before they solve the ten problems.

Games

Making Equations pp. 44–45

How Many More? pp. 46–48

Neighbors Count pp. 49–50

These open-ended and repeatable Games involve children in practicing many subtraction facts. *Making Equations* and *How Many More?* require children to record subtraction equations. In *Making Equations*, pairs collaborate to complete the activity. *Neighbors Count*, a challenging Game, encourages flexible thinking. All Games promote mental computation as children enhance their strategic thinking skills.

Independent Activities

Subtraction Squares pp. 51–52

Mystery Numbers Practice pp. 53–54

Making Equations Practice pp. 55–56

Each Independent Activity has two worksheets that require children to independently practice either the easier or the more difficult subtraction facts. *Subtraction Squares* provides lots of facts practice with ways to self-check. *Mystery Numbers Practice* reinforces the *Mystery Numbers* Sponge, while *Making Equations Practice* allows children to independently practice subtraction facts previously learned in a game format.

Ten Frames Subtraction

Topic: Subtraction Facts

Object: Visualize and verbalize differences.

Groups: Whole class or small group

Materials

- transparency of Ten Frames, p. 156
- counters (cubes are easily handled)

Tip For less-confident children, begin with smaller amounts to make totals of ten counters.

Directions

1. The leader displays eight counters in the upper Ten Frame on the Ten Frames transparency.

2. The leader asks children to identify the amount shown and asks, "How many more would you need to make 12?" When children respond, leader asks them to explain what they did mentally or visually to find the difference.

3. The leader displays the related subtraction fact that illustrates the problem 12 – 8. (When subtracting, the total and one of the two parts are known.)

4. The leader poses similar problems, starting with a display of eight counters, and asks children how many more are needed to make a variety of totals. For each of these totals, children help the leader identify the related subtraction fact.

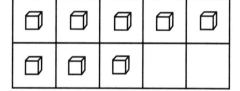

5. The leader then poses similar problems with a new starting amount.

   Example: Begin by displaying 6 counters and asking, "How many more are needed to have a total of 13?"

6. Remember to have children identify the related subtraction problem, orally or in writing, for each problem posed.

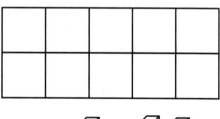

Making Connections

Promote reflection and make mathematical connections by asking:

- How do Ten Frames help you find differences for the subtraction facts?

Nimble with Numbers

Sponge

Mystery Numbers

Topic: Subtraction and Addition Facts

Object: Identify two numbers when given the sum and difference.

Groups: Whole class or small group

Materials

- transparency of *Mystery Numbers,* p. 38
- *Mystery Numbers* activity form (for each child), p. 38
- 4 pieces of 3-by-3-inch self-stick note paper (2 cut into triangle shapes using the sticky side as the base and 2 cut as squares)
- Digit Squares for each child, p. 149

Tip As children gain confidence, have them create Mystery Number Puzzles *for their classmates to solve. This is a more challenging activity if you begin with the subtraction clue first.*

Directions

1. The leader provides the visual clue shown at the right and announces, "The sum of my mystery numbers equals 10." $\square + \triangle = 10$

    (The transparency strip has the complete equation, 6 + 4 = 10, with the 6 covered by a square shape and the 4 covered by a triangle shape.)

2. After asking "What numbers could you use to complete this equation?" the leader displays each pair of numbers suggested.

3. Next the leader provides the visual clue shown at the right and states, "The difference of my two mystery numbers equals 2. Do any of our addition combinations fit both of these number sentences?" The leader clarifies by restating the two conditions: "The sum of my mystery numbers is 10, and the difference of my mystery numbers is 2. What are my mystery numbers?" $\square - \triangle = 2$

4. Children hold up two Digit Squares to identify the mystery numbers.

5. After children identify the mystery numbers, the leader asks, "If the two mystery numbers are . . . , which number does the triangle represent? Which number does the square represent?" (After children respond, the leader removes the shapes.)

6. Other possibilities for future mystery numbers:
    sum = 11 and difference = 3; sum = 11 and difference = 3;
    sum = 13 and difference = 5; and sum = 12 and difference = 2.

Making Connections

Promote reflection and make mathematical connections by asking:

- Why does the order of the mystery numbers matter?

Sponge

Mystery Numbers

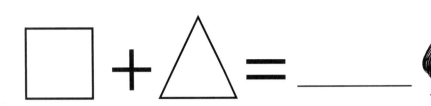

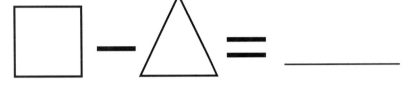

1 2 3 4 5 6 7 8 9

Number Trails

Topic: Subtraction and Addition Facts

Object: Identify sequences of touching numbers that can be added and subtracted to equal a given target number.

Groups: Whole class or small group

Materials

- transparency of *Number Trails*, p. 40
- overhead pens

Directions

1. The leader chooses and records a target number in the empty cell on the transparency.
2. Children take turns identifying a variety of trails that equal the target number. The trails must include subtraction and may include addition as well.
3. Encourage children to seek many trails.
4. The leader changes the target number. Children play additional rounds. (Numbers 0–6 work well as target numbers.)

Making Connections

Promote reflection and make mathematical connections by asking:

- What strategy helped you get to the target number?
- Were there certain numbers you avoided? Why?

Challenges

- Find a trail that uses more than six numbers.
- Create a trail that alternates subtraction and addition.
- Try to make a trail that equals 4 and passes through every number.

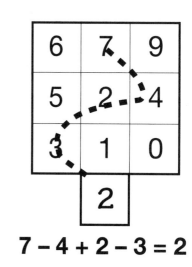

$7 - 4 + 2 - 3 = 2$

Sponge   Addition and Subtraction Facts

Number Trails

Date _____   Name _____

Just the Facts 7

 STOP Don't start yet! Star two problems that may have even answers.

1. 7 – 4 = _____   **2.** 10 – 6 = _____

3. 11 – 3 = _____   **4.** 14 – 7 = _____

5. 3 + _____ = 10

6. 10 – 3 = _____   **7.** 10   **8.** 12   **9.** 15   **10.** 16
                          – 5   – 4   – 8   – 9
                          ___   ___   ___   ___

 Go On What number is missing? 12, 10, 8, _____, 4, 2

------ ✂ ------

Date _____   Name _____

Just the Facts 8

 STOP Don't start yet! Star two problems that may have an answer less than 7.

1. 6 – 3 = _____   **2.** 9 – 5 = _____

3. 12 – 4 = _____   **4.** 13 – 6 = _____

5. 6 + _____ = 10

6. 10 – 6 = _____   **7.** 10   **8.** 12   **9.** 14   **10.** 16
                          – 2   – 6   – 6   – 8
                          ___   ___   ___   ___

 Go On If △ + □ = 10 and △ – □ = 2.

△ = _____   □ = _____

Skill Checks   Subtraction Facts   41

Date _____ Name _____

Just the Facts 9

 Don't start yet! Star two problems that may have odd answers.

1. 5 − 2 = _____   **2.** 10 − 7 = _____

3. 11 − 6 = _____   **4.** 15 − 9 = _____

5. 5 + _____ = 9

6. 9 − 5 = _____   **7.** 9   **8.** 11   **9.** 16   **10.** 15
                           −7   −2   −7   −6
                           ___   ___   ___   ___

 What number is missing? 20, 15, _____ , 5

Date _____ Name _____

Just the Facts 10

 Don't start yet! Star the problem in Row 1 that may have the smaller answer.

1. 5 − 4 = _____   **2.** 10 − 8 = _____

3. 12 − 5 = _____   **4.** 15 − 7 = _____

5. 9 + _____ = 12

6. 12 − 9 = _____   **7.** 9   **8.** 11   **9.** 18   **10.** 14
                           −6   −7   −9   −7
                           ___   ___   ___   ___

 If △ + □ = 12 and △ − □ = 4

□ = _____   △ = _____

42 Nimble with Numbers   Skill Checks

Date _____ Name _____

Just the Facts 11

STOP Don't start yet! Star the problem in Row 2 that may have an even answer.

1. 8 – 3 = _____　　　**2.** 9 – 1 = _____

3. 12 – 9 = _____　　**4.** 13 – 7 = _____

5. 6 + _____ = 9

6. 9 – 6 = _____

7. 10　　**8.** 11　　**9.** 17　　**10.** 13
　　 − 2　　　　 − 5　　　　 − 8　　　　 − 5
　　 ___　　　　 ___　　　　 ___　　　　 ___

Go On What number is missing? 18, 14, 10, _____, 2

- -

Date _____ Name _____

Just the Facts 12

STOP Don't start yet! Star the problems that may answers greater than 6.

1. 8 – 4 = _____　　**2.** 9 – 3 = _____

3. 12 – 7 = _____　　**4.** 13 – 9 = _____

5. 3 + _____ = 11

6. 11 – 3 = _____

7. 9　　**8.** 11　　**9.** 16　　**10.** 14
　　 − 2　　　　 − 3　　　　 − 9　　　　 − 5
　　 ___　　　　 ___　　　　 ___　　　　 ___

Go On If △ + □ = 9 and △ − □ = 3.

△ = _____　　□ = _____

Skill Checks　　　　　　　　　　　　　　　　Subtraction Facts　43

Making Equations

Topic: Subtraction and Addition Facts

Object: Make equations to equal 1 through 6.

Groups: Pairs

Materials for each pair
- *Making Equations* recording sheet, p. 45
- Digit Squares (digits 0–6 only), p. 149

Directions
1. Each pair collaborates to complete one recording sheet.
2. Each pair randomly places the Digit Squares 0–6 facedown. The pair selects and displays four of the Digit Squares.
3. Each pair uses three of the four displayed digits to form an equation. If the equation equals an amount shown on the recording sheet, the pair records the solution. If the equation equals an amount not shown, the pair rearranges the digits or selects a different combination of three Digit Squares until they find an equation that will fit.
4. Each pair repeats these steps until it has completed equations for each of the six amounts shown on the recording sheet.

   Option: When pairs gain confidence with this activity, suggest pair members alternate turns to complete the recording sheet.

Tip Eventually children could have their own recording sheets and alternate turns to see who can complete a recording sheet in the fewest turns.

Making Connections
Promote reflection and make mathematical connections by asking:
- Which solutions were not possible?
- What type of numbers do you need to make an even answer?

	+/−		+/−		= 1
	+/−		+/−		= 2
6	⊕	2	⊖	5	= 3
	+/−		+/−		= 4
	+/−		+/−		= 5
	+/−		+/−		= 6

44   Nimble with Numbers   Game

Making Equations

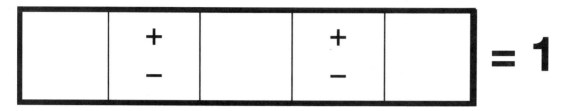 = 1

 = 2

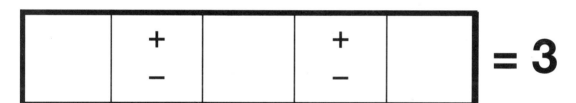 = 3

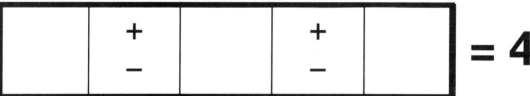 = 4

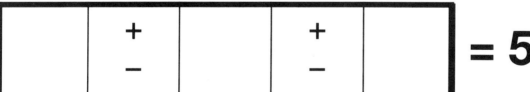 = 5

 = 6

Game — Subtraction Facts

How Many More?

Topic: Subtraction Facts

Object: Cover three numbers in a row.

Groups: Pair players or 2 players

Materials for each group
- *How Many More to Make 10?* gameboard, p. 47
- markers (different kind for each pair)
- Number Cube (1–6), p. 151
- paper for recording equations

Directions

1. The first pair rolls the number cube and decides how many more are needed to make 10. Pair places a marker on a gameboard number that represents the missing amount.

   Example: If 3 is rolled, 7 is needed to make 10. The pair covers one of the 7s on the gameboard.

2. Pairs are required to say aloud and record the related subtraction fact for each turn on a separate sheet of paper.

   Example: If 3 is rolled, a pair records 10 – 3 = 7.

3. Pairs alternate turns and follow this procedure until one pair wins by placing three markers in a row horizontally, vertically, or diagonally.

   Option: To provide practice with more difficult facts, select *Gameboard A* or *Gameboard B,* both of which are designed to be used with the 4–9 number cube (or spinner). *Gameboard A* works well with subtraction facts 13 through 15, while *Gameboard B* is designed for subtraction facts 16 through 18.

Making Connections
Promote reflection and make mathematical connections by asking:
- What strategies did you use in placing your markers?
- Were some numbers easier to cover than others?
- Was it difficult to block your opponent? Why or why not?

Tip As an easy replacement for number cubes, consider using the 1–6 Digit Cards or Dot Pattern Cards randomly mixed and stacked facedown. Gameboards A and B will require the 4–9 Digit Cards or Digit Squares.

14				
5	8	4	7	11
7	6	9	8	7
11	9	8	6	9
6	7	9	8	10
8	5	10	4	6

How Many More to Make 12?

8	7	10	7
6	8	9	11
10	9	10	8
7	11	6	9

$3 + ? = 12$

How Many More to Make 10?

4	6	9	7
8	7	5	4
6	4	9	5
7	5	6	8

$3 + ? = 10$

How Many More? A   How Many More? B

(8 + ? = 14)

(8 + ? = 17)

5	8	4	7	11
7	6	9	8	7
11	9	8	6	9
6	7	9	8	10
8	5	10	4	6

14	7	10	8	14
13	10	9	11	12
9	11	12	10	11
12	9	11	12	7
8	11	10	13	9

Nimble with Numbers

Game

Neighbors Count

Topic: Subtraction and Addition Facts

Object: Score the highest total.

Groups: Pair players or 2 to 4 players

Materials for each group
- *Neighbors Count* gameboard, p. 50
- 3 Number Cubes (1–6), p. 151
- markers (8 for each pair)
- paper for keeping score

Tip Students soon discover it's disadvantageous to go first, so you might want to announce how groups should determine order of play.

Directions:

1. The first player or pair rolls three cubes, then adds and/or subtracts the three numbers shown. After the equation is stated, the resulting answer is located and covered with a marker.

   Example: With 1, 2, and 6 you might say "$1 + 2 + 6 = 9$" and cover 9 on the gameboard. Other possibilities are: "$1 + 6 - 2 = 5$" and cover 5, or "$6 - 2 - 1 = 3$" and cover 3.

2. The next player rolls the three cubes and uses the three rolled numbers to find a new answer. A point is scored if the player can cover an uncovered number that shares a side with an already covered number.

3. Players continue to follow this process. As more numbers are covered, players earn more points, since one point is scored for each adjacent covered number.

   Example: If 1, 2, 5, and 7 are covered and you are able to make 6, you receive 3 points.

4. Each player keeps a running total, recording a score after each turn. Sometimes a player will be unable to make any uncovered number and must pass for that turn. The game ends after all numbers are covered or after three consecutive passes by one player. The player with the highest cumulative score wins.

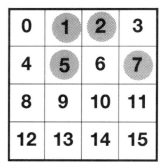

$6 - 2 - 1 = 3$

Making Connections

Promote reflection and make mathematical connections by asking:
- Which kinds of rolls do you prefer? Please explain.
- What strategies worked best to give you high totals on your turns?

Neighbors Count

0	1	2	3
4	5	6	7
8	9	10	11
12	13	14	15

Subtraction Squares I

Subtract each row and column to fill in the missing answers.

Sample

9	5	4
3	1	2
6	4	2

9	4	
6	3	

6	4	
5	3	

8	6	
5	4	

7	3	
5	1	

9	2	
7	1	

5	4	
3	2	

7	3	
5	1	

8	5	
3	2	

Independent Activity — Subtraction Facts 51

Date _____   Name _____

Subtraction Squares II

Subtract each row and column to fill in the missing answers.

Sample

10	5	*5*
3	1	*2*
7	*4*	*3*

10	6
5	3

12	6
5	2

11	7
5	1

13	8
7	6

15	7
6	3

12	8
7	5

11	6
9	5

14	8
6	3

Nimble with Numbers   Independent Activity

Date _____ Name _____

Mystery Numbers Practice 1

☐ + △ = 11
☐ − △ = 1
☐ = _____
△ = _____

☐ + △ = 12
☐ − △ = 4
☐ = _____
△ = _____

☐ + △ = 12
☐ − △ = 2
☐ = _____
△ = _____

☐ + △ = 10
☐ − △ = 2
☐ = _____
△ = _____

☐ + △ = 10
☐ − △ = 6
☐ = _____
△ = _____

☐ + △ = 11
☐ − △ = 3
☐ = _____
△ = _____

Independent Activity Subtraction Facts

Date _____ Name _____

Mystery Numbers Practice II

Clue: All squares on this page cover 2-digit numbers.

☐ + △ = 17
☐ − △ = 5
☐ = _____
△ = _____

☐ + △ = 18
☐ − △ = 10
☐ = _____
△ = _____

☐ + △ = 16
☐ − △ = 8
☐ = _____
△ = _____

☐ + △ = 17
☐ − △ = 9
☐ = _____
△ = _____

☐ + △ = 17
☐ − △ = 7
☐ = _____
△ = _____

☐ + △ = 18
☐ − △ = 8
☐ = _____
△ = _____

Nimble with Numbers **Independent Activity**

Making Equations Practice I

Use 1, 2, and 4 to complete these number sentences.

1. ☐ − ☐ − ☐ = 1
2. ☐ + ☐ − ☐ = 3

Use 1, 3, and 6 to complete these number sentences.

3. ☐ − ☐ + ☐ = 4
4. ☐ − ☐ − ☐ = 2

Use 1, 3, and 5 to complete these number sentences.

5. ☐ + ☐ − ☐ = 7
6. ☐ + ☐ − ☐ = 3

Use + or − signs to complete these number sentences.

7. 4 ☐ 3 ☐ 2 = 5
8. 4 ☐ 3 ☐ 2 = 3
9. 5 ☐ 3 ☐ 1 = 7
10. 5 ☐ 3 ☐ 1 = 3

Independent Activity

Subtraction Facts 55

Making Equations Practice II

Use 4, 5, and 8 to complete these number sentences.

1. ☐ − ☐ + ☐ = 7
2. ☐ + ☐ − ☐ = 1
3. ☐ − ☐ + ☐ = 9

Use 7, 8, and 9 to complete these number sentences.

4. ☐ + ☐ − ☐ = 6
5. ☐ + ☐ − ☐ = 8
6. ☐ − ☐ + ☐ = 10

Use + or − signs to complete these number sentences.

7. 9 ☐ 4 ☐ 5 = 10
8. 9 ☐ 4 ☐ 5 = 8
9. 9 ☐ 4 ☐ 5 = 18

Show how many different equations you can make using 9, 6, and 2.

Nimble with Numbers — Independent Activity

Money

Assumptions Money concepts have been previously taught with an emphasis on understanding and an effort to enhance number sense. Concrete and visual models, such as coins and play money, have been used extensively. Children have practiced finding the value of many coin combinations.

Section Overview and Suggestions

Sponges

Counting Coins p. 58

Guess My Coins p. 59

What's My Change? pp. 60–61

These open-ended, repeatable warm-ups emphasize mental combining of coin values. Frequent use of *Counting Coins* provides a foundation for success with future activities in this section. *Guess My Coins* and *What's My Change?* accommodate varying levels of difficulty.

Skill Checks

Sensible Cents 1–3 pp. 62–64

The Skill Checks provide a way to see children's improvement in working with money. Children should be allowed access to play coins. Have children respond to STOP before they solve the ten problems.

Games

Race to 50¢ pp. 65–66

Pennies or Nickels? p. 67

Who Buys? pp. 68–69

These repeatable Games actively involve children in mentally combining coins. *Race to 50¢* requires children to make appropriate money exchanges. *Pennies or Nickels?* develops number sense. Adequate exposure to the *What's My Change?* Sponge should ensure children success with the challenging *Who Buys?* Game that involves strategic thinking. All three engaging Games are ones parents will enjoy playing repeatedly with their children.

Independent Activities

Buying Snacks p. 70

Find the Combination p. 71

What's My Change? p. 72

Each Independent Activity engages children in mentally combining coins and strengthens skills previously developed while experiencing the Sponges and Games. Children may appreciate access to play coins. *Find the Combination* asks children to make combinations using a specified number of coins. Each Independent Activity presents real-life situations like requiring children to make change in *What's My Change?*.

Counting Coins

Topic: Adding Coins

Object: Count coin combinations with skip counting by 10s, 5s, 1s, and/or 25s.

Groups: Whole class or small group

Materials
- transparent Coins, p. 147

Directions

1. The leader announces, "Let's count groups of coins. Count aloud as each coin is placed on the overhead. When dimes are placed on the overhead, what counting pattern will you use?" (counting by tens)

2. The leader places four dimes, one at a time, on the overhead. As each dime is placed, children count by tens, "10. . . 20. . . 30. . . 40."

3. The leader changes the counting sequence by placing three nickels, one at a time, on the overhead. As each nickel is placed, children continue counting, but now count by fives, "45. . . 50. . . 55."

4. The leader states, "The total value of these coins is 55¢. Let's count this same amount again." Place four dimes and three nickels on the overhead again, and repeat the counting sequence, "10. . . 20. . . 30. . . 40. . . 45. . . 50. . . 55."

5. The leader follows the same procedure but places three quarters and four pennies on the overhead, one at a time. The counting sequence now is, "25. . . 50. . . 75. . . 76. . . 77. . . 78. . . 79."

6. After counting with two types of coins, the leader has children count totals using three different types of coins. As the number and types of coins are varied, the steps are repeated.

Making Connections

Promote reflection and make mathematical connections by asking:
- If you had a handful of coins, how might you count coins more easily?
- What mathematics are you doing in this activity?

Tip If coins are counted with larger values first, children usually experience more success (quarters, then dimes, nickels, or pennies). As children gain confidence with this activity, switch the order of coins. For example, place three dimes, two pennies, four nickels, and two dimes, counting "10. . . 20. . . 30. . . 31. . . 32. . . 37. . . 42. . . 47. . . 52. . . 62. . . 72."

Guess My Coins

Topic: Mental Addition of Coins

Object: Determine exact coin combinations.

Groups: Whole class or small group

Materials

- prepared coin combination containers (labeled with the total value)
- play coins for children

Tip Make this activity more interesting by encouraging less common questions such as, "Do you have coins with bearded presidents?" or "Do you have any copper coins?"

Directions

1. The leader shakes one container of coins and announces the value of the coins.

   Example: "These coins total 25¢."

2. Children use play coins to display one possible combination equaling the amount announced.

3. Children ask the leader questions that can be answered with "yes" or "no."

   Example: "Do you have more than three coins?"

4. The leader responds appropriately with "yes" or "no."

5. Based on the information given by the leader, some children may revise their displayed coin combinations.

6. Children continue to ask questions until most children are displaying the exact combination of coins in the container.

7. As children gain skill with repeated experience using different coin values, the leader varies the activity by announcing only the number of coins in the container, not the total value.

   Example: "I have 4 coins in my container."

8. Children ask "yes or no" questions and display the coins accordingly. With enough questions, children eventually identify the coins and the value of the coins in the container.

Making Connections

Promote reflection and make mathematical connections by asking:
- How did you know which questions to ask?
- What mathematics did you use to find your solution?

Sponge

What's My Change?

Topic: Mental Addition and Subtraction of Coins

Object: Determine coin-combination payment and exact change.

Groups: Whole class or small group

Materials

- transparency of *What's My Change?* (illustrated items cut apart), p. 61
- transparent Coins, p. 147
- play coins for children (*optional*)

Directions

1. Leader displays one selected item, asks children to suggest an appropriate value, and records the amount on the price tag.

   Example: The notepad is selected. The class suggests 15¢ as a price and writes "15¢" on the price tag.

2. Leader explains the buyer never has exact change and asks children to identify one possible combination of coins to pay for the displayed item. A volunteer places the suggested combination of transparent coins on the money bag.

3. Leader asks children to identify an exact coin combination the buyer might receive as change. Volunteers share possibilities by counting on from the purchase price and placing appropriate coins on the hand.

   Example: Students decide to pay with a quarter and place a quarter in the bag. Two transparent nickels are placed on the hand as children count, "15 cents . . . 20 cents . . . 25 cents."

4. After solutions are shared, children suggest another way the payment might be made and repeat steps 2 and 3.

5. Students select other items, determine prices, and display coin combinations for the purchase price and change received. Encourage children to suggest prices other than multiples of 5¢.

6. Eventually, the leader places two items in the purchase box and displays a second price tag. Children repeat the process.

Tips Have children find out how class-generated prices compare to prices in local stores. Suggested price ranges for the illustrated items: Notepad 15¢–50¢; Pencil 10¢–30¢; Crayons 39¢–99¢; Sticker 3¢–20¢, Eraser 5¢–35¢; Pencil Sharpener 19¢–75¢.

Making Connections

Promote reflection and make mathematical connections by asking:

- What strategy did you use to determine the exact change?

What's My Change?

PAYMENT

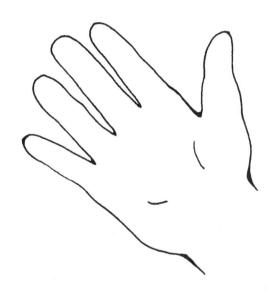
CHANGE

 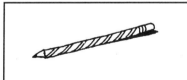 

  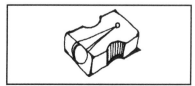

Date _____   Name _____

Sensible Cents 1

🛑 **STOP** Don't start yet! Star two problems that may have coins worth more than 35¢.

1. 21¢

2. 36¢

3.  = _____ ¢

4.  = _____ ¢

5. Use 2 coins to make 20¢.

○ ○

6. Use 3 coins to make 20¢.

○ ○ ○

7–8. Find the change.

Price	Payment	Change
24¢		_____ ¢
45¢		_____ ¢

9–10. Show 40¢ two ways.

▶ **Go On** These shapes have values of △1¢ ○5¢ ☐10¢ ☐25¢ .
Draw a picture worth 60¢.

62 Nimble with Numbers   Skill Check

Date _____   Name _____

Sensible Cents 2

 Don't start yet! Star two problems that may have coins worth less than 40¢.

1. 25¢

2. 55¢

3.  = _____ ¢

4.  = _____ ¢

5. Use 2 coins to make 15¢.

   ◯ ◯

6. Use 3 coins to make 51¢.

   ◯ ◯ ◯

7–8. Find the change.

Price	Payment	Change
29¢		_____ ¢
40¢		_____ ¢

9–10. Show 32¢ two ways.

 Use the fewest number of coins to make 45¢. Explain how you know.

Skill Check   Money   63

Date _____ Name _____

Sensible Cents 3

 Don't start yet! Star two problems that may have coins worth more than 55¢.

1. 45¢

2. 27¢

3.  = _____ ¢

4.  = _____ ¢

5. Use 2 coins to make 26¢.
   ◯ ◯

6. Use 3 coins to make 60¢.
   ◯ ◯ ◯

7–8. Find the change.

Price	Payment	Change
31¢		_____ ¢
53¢		_____ ¢

9–10. Show 49¢ two ways.

 These shapes have values of Draw a picture worth 83¢. △ 1¢   ◯ 5¢   ☐ 10¢   ☐ 25¢

Race to 50¢

Topic: Exchanging and Adding Pennies, Nickels, Dimes, and Quarters

Object: Reach 50¢.

Groups: 2 players or pair players

Materials for each group
- *Race to 50¢* gameboard for each player, p. 66
- 2 Number Cubes (1–6), p. 151
- 20 markers
- play coins: pennies, nickels, dimes, and quarters *(optional)*

Directions

1. The first player rolls two number cubes. The number rolled indicates the number of pennies awarded for that turn. The player covers the rolled amount on his or her gameboard and states the accumulated value. After accumulating 5 pennies, a player must exchange them for a nickel.

2. The second player rolls the number cubes, indicates the value of the roll on his or her gameboard, and states the accumulated value.

3. Players continue to alternate turns and follow the same procedure. Players must exchange coins when appropriate (five pennies for a nickel, two nickels for a dime, two dimes and a nickel for a quarter). After accumulating a quarter, a player places a marker on the quarter; that player is halfway to a winning round. Players win when they have accumulated two quarters. If players have the same number of turns, it is possible both players could win.

4. Since exchanging coins is worthwhile practice, players are encouraged to play additional rounds.

Tips Reinforce subtraction and making change by reversing the rules. Begin with 50¢ (two quarters) and "Race to Zero." Some children might find it helpful to record their turns and keep track of their totals.

Making Connections
Promote reflection and make mathematical connections by asking:
- If you were to redesign the number cube, how would you change it? Why?

Race to 50¢

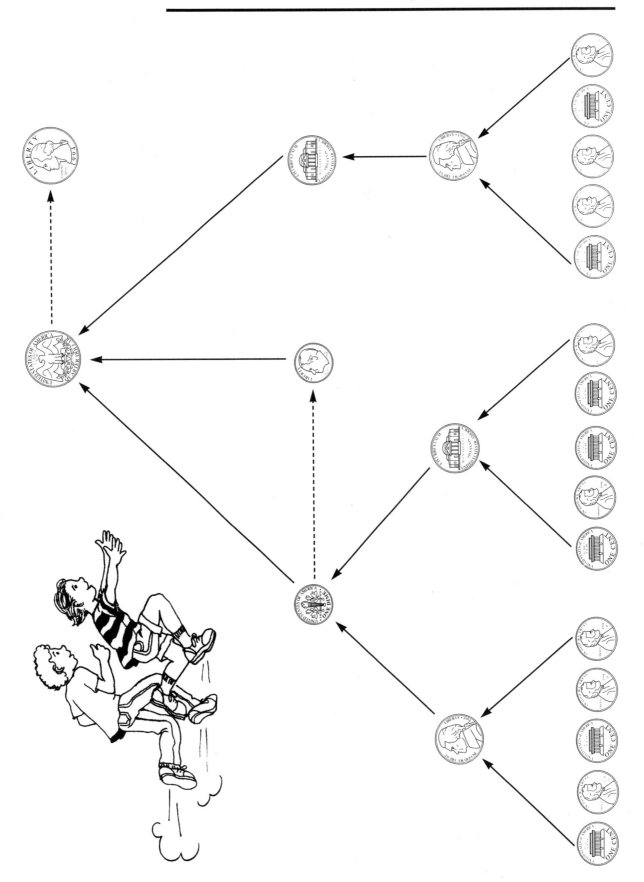

Pennies or Nickels?

Topic: Adding Nickels and Pennies

Object: Reach close to 25¢ in three rolls.

Groups: Pair players or 2 players

Materials for each group
- Number Cubes (1–6), p. 151
- play coins: nickels and pennies

Directions

1. The first pair rolls the number cubes. The number rolled indicates how many nickels or pennies the pair takes, not a combination of both coin amounts. The pair decides whether to take pennies or nickels and announces the total amount.

Example: "We rolled a three. We'll take three nickels. We have 15¢."

2. The second pair rolls and follows the same procedure. When appropriate, pairs may exchange five pennies for a nickel.

3. The first pair rolls again, decides whether nickels or pennies are preferred, takes that number of nickels or pennies, and adds it to the existing amount.

Example: "We had 15¢. We rolled a six. We'll take six pennies. 15¢ plus 6¢ equals 21¢. We now have 21¢."

4. Pairs alternate turns until each pair has rolled three times. After three rolls, each pair totals its coins. The pair closest to 25¢ wins.

Example: First pair: "We have 22¢. We are 3 away from 25¢."

Second pair: "We have 27¢. We are 2 away from 25¢. We win."

Making Connections

Promote reflection and make mathematical connections by asking:
- How did you decide whether to take pennies or nickels on each turn?
- How would you play differently in future games?

Tips Increase the difficulty of this game by allowing more rolls and a greater target amount. "Roll the number cubes six times and try for an amount close to 50¢." Students can also vary the value of coins selected by using dimes and pennies or dimes and nickels.

Who Buys?

Topic: Mental Addition of Coins

Object: Add coins to total a specified amount.

Groups: Pair players or 2 players

Materials for each group
- *Who Buys?* gameboard, p. 69
- 16 markers
- play coins: pennies, nickels, dimes, and quarters *(optional)*

Directions

1. Leader designates which item is to be purchased (target amount). Pairs place a marker on the indicated item.
2. The first pair places a marker on one of the coins and states the amount.
3. Pairs alternate turns by placing one marker on any uncovered coin and stating the new total.

   Example: If a dime and nickel are covered, the current total is 15 cents. If the next pair covers a nickel, the pair states "20 cents" as the new total.

4. The first pair to reach and state the purchase price exactly wins. If a pair goes over the target amount, the other pair wins that round.
5. Additional rounds can be played to purchase the same item, or a new item can be designated at the beginning of the game.

Making Connections
Promote reflection and make mathematical connections by asking:
- What strategy helped you get close to the purchase amount?

Tip Encourage children to play additional rounds seeking the same specified amount to see patterns and discover winning strategies. Less-experienced children might benefit from using price tags with prices in multiples of 5 cents.

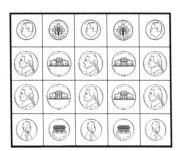

Who Buys?

Game

Money 69

Buying Snacks

Mark the coins you need to buy each item.

#	Item	Price	Coins
1.	apple	45¢	
2.	popcorn	70¢	
3.	raisins	42¢	
4.	crackers	35¢	
5.	 granola bar	51¢	
6.	fruit roll	60¢	
7.	nuts	85¢	
8.	beef jerky	38¢	

Independent Activity

Date _____ Name _____

Find the Combination

  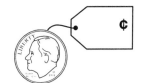 

Find a coin combination to match the amount on the price tag. Use the exact number of coins indicated by the circles.

1.
27¢ (25¢) (1¢) (1¢)

2.
16¢ ○ ○ ○

3.
36¢ ○ ○ ○

4.
40¢ ○ ○ ○

5.
17¢ ○ ○ ○ ○

6.
26¢ ○ ○ ○ ○

7.
25¢ ○ ○ ○ ○

8.
21¢ ○ ○ ○ ○

9.
30¢ ○ ○ ○ ○

10.
35¢ ○ ○ ○ ○

Independent Activity Money 71

Date _____ Name _____

What's My Change?

Find your change if you buy each item.

Item	Payment		Change
1. jawbreaker 22¢	(quarter)	25 ¢	3 ¢
2. mints 27¢	(dime, dime, dime)	____ ¢	____ ¢
3. gum 38¢	(quarter, dime, dime)	____ ¢	____ ¢
4. lollipop 29¢	(quarter, quarter)	____ ¢	____ ¢
5. licorice 31¢	(quarter, dime, nickel)	____ ¢	____ ¢
6. jelly beans 62¢	(quarter, quarter, dime, dime)	____ ¢	____ ¢
7. candy bar 55¢	(quarter, quarter, quarter)	____ ¢	____ ¢

Nimble with Numbers — Independent Activity

Place Value

Assumptions Place value concepts have been taught for understanding with an emphasis on developing number sense. Concrete and visual models such as Ten Frames, 100 charts, 99 charts, and money, have been used extensively.

Section Overview and Suggestions

Sponges

100 Chart Pieces pp. 74–75

Arrow Math p. 76

Place Value Paths pp. 77–78

Where Will It Fit? pp. 79–80

All the Sponges help children develop memory of the 100 chart, a valuable visual tool. *Place Value Paths* and *Where Will It Fit?* are engaging, enjoyable, repeatable warm-ups that develop number sense and strategic thinking.

Skill Checks

What's in That Place? 1–6 pp. 81–83

The Skill Checks provide a way to help parents, children, and you see children's improvement with place value. Remember to have all children respond to STOP before they solve the ten problems.

Games

100 Chart Cover p. 84

100 Chart Paths pp. 85

Will It Fit? pp. 86–87

Where? pp. 88–89

To ensure success with these strategic Games, which further the development of number sense, allow repeated experience with the Sponges. All of these Games encourage planning ahead and the thoughtful placement of digits. The challenging Games of *100 Chart Paths* and *Where?* require a good undertanding of place value.

Independent Activities

What Numbers Are Missing? pp. 90–92

Staircase Number Puzzles pp. 93–94

Creating Numbers pp. 95–96

What Numbers are Missing? and *Staircase Number Puzzles* engage children in visualizing numbers on the 100 Chart. *Creating Numbers* builds number sense and enhances logical thinking. Children could use Digit Squares when solving *Creating Numbers* puzzles. To extend the use of all these Independent Activities, encourage children to create similar activities for classmates to solve.

100 Chart Pieces

Topic: Number Relationships

Object: Determine unknown numbers by visualizing the 100 Chart.

Groups: Whole class or small group

Materials

- transparency of *100 Chart Pieces*, p. 75 (cut apart)
- 100 Chart for each child, p. 146
- marker for each child

Tip As children become competent, conduct this warm-up without the use of a 100 Chart.

Directions

1. The leader displays one of the simpler 100-chart pieces and writes a number in one of the cells.
2. The leader places a marker in one of the remaining empty cells.
3. Students try to determine what number belongs in the marked empty cell. They indicate their response by covering that number on their 100 Chart.
4. The leader continues to provide more challenging clues, and the players identify the unknown numbers with their one marker and the 100 Chart.
5. As this warm-up proceeds, the leader asks children to explain how they figured out the number. It's helpful to have different approaches shared.

Option: Increase the difficulty by having children determine the answer before referring to the 100 Chart (or extend the activity with a 200 Chart).

Making Connections

Promote reflection and make mathematical connections by asking:
- What patterns helped you identify the unknown number?

100 Chart Pieces

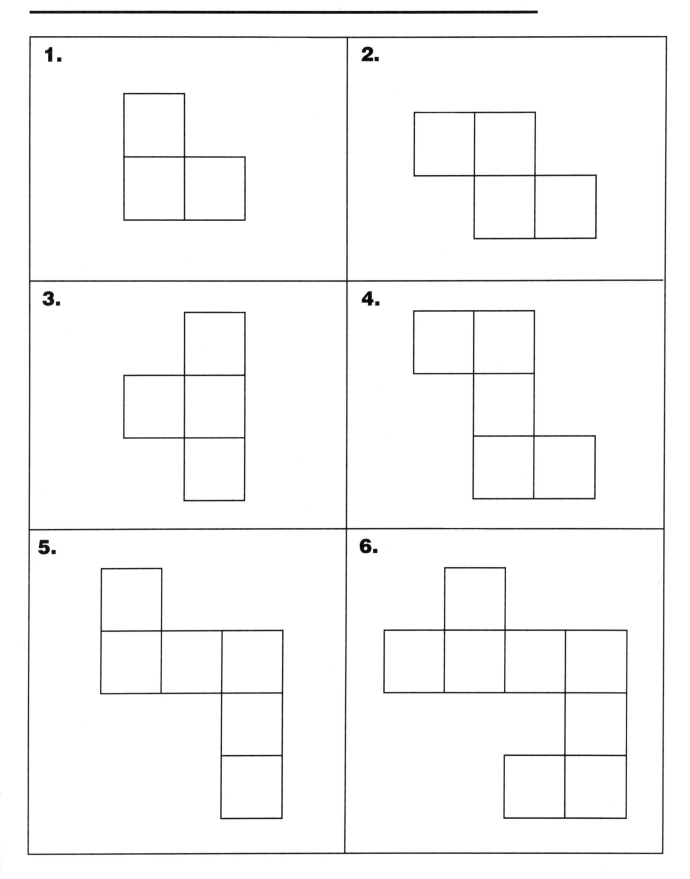

Arrow Math

Topic: Number Sense to 100

Object: Identify a coded number.

Groups: Whole class or small group

Materials

- 100 Chart (one for each child), p. 146
- marker (one for each child)
- chalkboard or overhead projector

Directions

1. Announce, "I'm thinking of a number on the 100 Chart. 16 ← is a clue to identify my number. Who's willing to guess what my number might be?" (Acknowledge that 16 ← identifies 15.)

2. Have children try to identify other numbers with clues.

    36 ← (identifies 35)
    47 → → (identifies 49)
    27 ↓ (identifies 37)
    34 ↑ (identifies 24)
    18 ↓ ↓ (identifies 38)
    46 ↓ ← ← (identifies 54)
    25 ↑ ← (identifies 14)

3. Have children create and display clues for classmates to solve.

4. When children appear to respond quickly and confidently, challenge them to identify coded numbers without the aid of the 100 Chart.

Making Connections

Promote reflection and make mathematical connections by asking:

- What happens to a number followed by a ↓?
- What other patterns have you noticed?

Place Value Paths

Topic: Number Sense

Object: Record numbers in an ascending sequence.

Groups: Whole class or small group

Materials

- 2 sets of Digit Cards, p. 148
- *Place Value Paths* recording sheet for each child, p. 78
- 100 Chart, p. 146 *(optional)*

Directions

1. The leader mixes the two sets of Digit Cards together and stacks them facedown. The leader then draws two cards and announces the digits to the class and asks, "What numbers can be formed using these digits?"

Example: If 8 and 3 are drawn, the number choices are 38 or 83.

2. Each child selects one of the choices. The leader reminds children that ultimately their paths must contain six two-digit numbers that are in order from smallest to largest. Initially, the leader helps children consider an area where their numbers fit appropriately between 10 and 100.

3. Each child independently records his or her number choice in one of the cells along the place value path. (Drawn cards are set aside.)

4. If a child cannot place either of the possible numbers in any of his or her remaining cells, nothing is recorded.

5. After six draws, the leader asks if anyone has completed an entire place value path. Draws continue until the majority of children have completed paths. Children compare their final results.

Making Connections

Promote reflection and make mathematical connections by asking:
- How did your path differ from others?
- How did you decide where to place your numbers?

Tip Some children appreciate the use of a 100 chart to help determine placement of numbers.

Place Value Paths

Where Will It Fit?

Topic: Place Value

Object: Put three numbers in a row, following the correct number sequence.

Groups: Whole class or small group

Materials

- transparency of *Where Will It Fit?* recording form, p. 80
- 2 overhead pens (different colors)
- Number Cube (1–6), p. 151
- special Number Cube (3–8), p. 151
- 100 Chart, p. 146 *(optional)*

Tip Some children appreciate the use of the 100 chart, p. 146, to help determine placement of selected numbers.

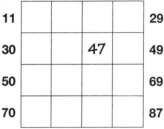

Directions

1. Explore with children which numbers are possible if the number cube and the special number cube (3–8) are rolled and used as digits for two-digit numbers. Ask, "Which two-digit numbers between 10 and 40 are *not* possible?"
2. Divide the group into two teams. Explain the object of the game, emphasizing the importance of placing the numbers in the correct row and ordering each row of numbers from smallest to largest.
3. A player on the first team rolls two cubes and announces the possible two-digit numbers. The team decides which number to use and tells the leader where to record the number. The leader records the selected number in the team's color.

   Example: If 4 and 7 are rolled, the team can select and appropriately record 47 in the second row or record 74 in the fourth row.

4. The other team takes a turn as explained in step 3.
5. Teams alternate turns until one team records three of its numbers in a row, horizontally, vertically, or diagonally. When a team's roll produces numbers that cannot be placed, that team passes.

Making Connections

Promote reflection and make mathematical connections by asking:

- What did you try that didn't work?
- What strategies helped you line up numbers next to each other?

Sponge

Where Will It Fit?

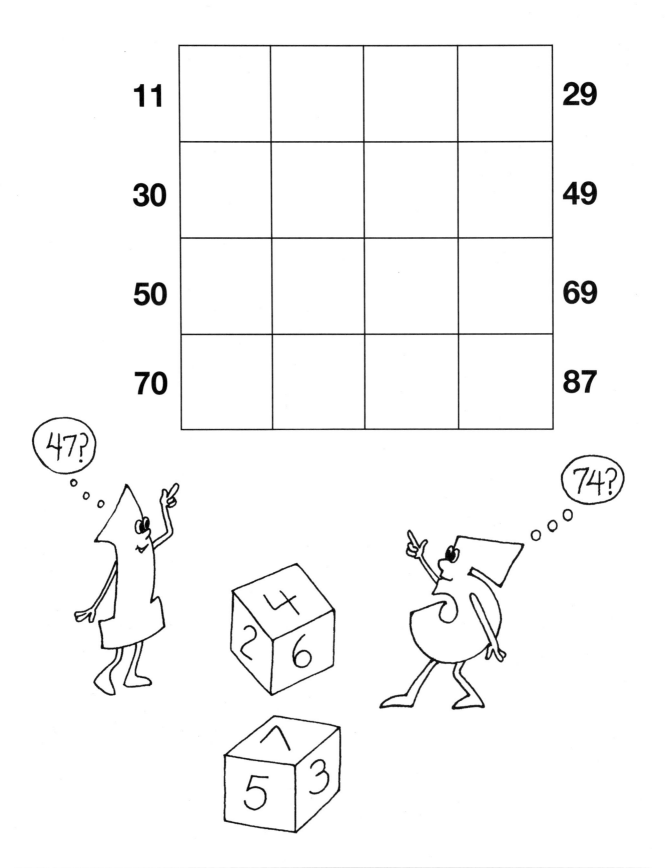

Date _____   Name _____

What's in That Place? 1

STOP Don't start yet! Star a problem that may have an answer larger than 45.

Fill in the missing numbers from the 100 Chart

1. [ ][33]
    [ ][ ]

2. [21][ ]
    [ ][ ]

3. Order these numbers from smallest to largest: 42, 57, 24 _____

4. thirty-seven _____   5. fifty-six _____   6. 2 tens less than 45 _____

Use two of the digits 3, 4, and 5 to form:

7. even number between 42 and 56 _____   8. odd number greater than 50 _____

9. even number less than 50 _____   10. number between 50 and 75 _____

Go On What numbers come next? 22, 32, 42, _____ , _____ .

✂ ···

Date _____   Name _____

What's in That Place? 2

STOP Don't start yet! Star the problems in Row 3 that may have odd answers.

Fill in the missing numbers from the 100 Chart.

1. [ ][ ]
    [64][ ]

2. [ ]
    [ ]
    [42]

3. Order these numbers from smallest to largest: 63, 46, 50 _____

4. forty-five _____   5. seventy-two _____   6. 3 tens more than 27 _____

Use two of the digits 2, 4, and 7 to form:

7. even number between 40 and 50 _____   8. odd number greater than 30 _____

9. even number less than 40 _____   10. number between 72 and 90 _____

Go On What is closest to 23? 20, 31, 24, 19 _____ Explain your answer.

Skill Checks   Place Value   81

Date _____ Name _____

What's in That Place? 3

 Don't start yet! Star the problems in Row 3 that may have the smallest answer.

Fill in the missing numbers from the 100 Chart:

1. 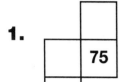 (75)

2. (55)

3. Order these numbers from smallest to largest: 84, 64, 62 _____

4. forty-eight _____ **5.** sixty-three _____ **6.** 2 tens less than 62 _____

Use two of the digits 5, 6, and 7 to form:

7. odd number between 70 and 80 _____ **8.** even number greater than 60 _____

9. odd number less than 60 _____ **10.** number between 50 and 70 _____

 What other number fits? | 46 | | |
                                              | | 40 | 42 |

Date _____ Name _____

What's in That Place? 4

 Don't start yet! Star the problem that may have the largest answer.

Fill in the missing numbers from the 100 Chart:

1. (46)

2. (28)

3. Order these numbers from smallest to largest: 27, 43, 36 _____

4. twenty-nine _____ **5.** fifty-four _____ **6.** 3 tens more than 21 _____

Use two of the digits 1, 2, and 4 to form:

7. even number between 20 and 30 _____ **8.** odd number less than 30 _____

9. even number greater than 30 _____ **10.** even number between 40 and 50 _____

 Write a two-digit number whose digits add up to 8. _____

82 *Nimble with Numbers*   **Skill Checks**

Date _____ Name _____

What's in That Place? 5

STOP Don't start yet! Star two problems that may have answers greater than 50.

Fill in the missing numbers from the 100 Chart:

1.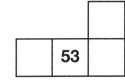
2. [chart with 68]

3. Order these numbers from smallest to largest: 52, 25, 18 _____

4. fifteen _____   **5.** eighty-two _____   **6.** 2 tens less than 74 _____

Use two of the digits 4, 8, and 9 to form:

7. odd number between 80 and 90 _____   **8.** number greater than 95 _____

9. even number less than 60 _____   **10.** even number between 80 and 90 _____

Go On Which is closest to 53? 51, 42, 54, 57 _____ Explain your answer.

Date _____ Name _____

What's in That Place? 6

STOP Don't start yet! Star the problem in Row 3 that may have the largest answer.

Fill in the missing numbers from the 100 Chart:

1.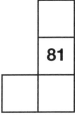
2.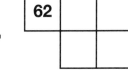

3. Order these numbers from smallest to largest: 43, 39, 52 _____

4. thirty-six _____   **5.** eighty-nine _____   **6.** 3 tens more than 18 _____

Use two of the digits 2, 6, and 7 to form:

7. even number between 20 and 30 _____   **8.** odd number less than 30 _____

9. even number greater than 30 _____   **10.** even number between 40 and 50 _____

Go On Write a two-digit number whose digits add up to 8.

Independent Activity   Place Value 83

100 Chart Cover

Topic: Place Value

Object: Place one marker in each row.

Groups: Pair players or 2–4 players

Materials for each group

- 100 Chart (one for each player), p. 146
- markers
- 2 sets of Digit Cards, p. 148

Directions

1. A player mixes two sets of Digit Cards and stacks them facedown.
2. Each pair (player) takes two Digit Cards and forms a two-digit number. The formed number is stated and covered on the 100 Chart.
3. Play continues until one or both pairs place at least one marker in each row. Keep in mind the zero will need to be placed in the tens column to form most numbers in the first row.

Making Connections

Promote reflection and make mathematical connections by asking:

- What strategy helped you cover a number in every row?
- How could this game be made more challenging?

| 4 | 6 |

1	2	3	4	5	6	7	8	9	10
11	12	13	14	15	16	17	18	19	20
21	22	23	24	25	26	27	28	29	30
31	32	33	34	35	36	37	38	39	40
41	42	43	44	45		47	48	49	50
51	52	53	54	55	56	57	58	59	60
61	62	63	64	65	66	67	68	69	70
71	72	73	74	75	76	77	78	79	80
81	82	83	84	85	86	87	88	89	90
91	92	93	94	95	96	97	98	99	100

100 Chart Paths

Topic: Place Value

Object: Form a continuous path from one side to the other.

Groups: Pair players or 2 players

Materials for each group

- 100 Chart (gameboard), p. 146
- different-colored markers for each pair
- 2 sets of Digit Cards, p. 148

Tip To increase possibilities and enhance strategic thinking, have pairs draw three cards and use two of them to form a number.

Directions:

1. After the Digit Cards are mixed and stacked, the first pair draws two Digit Cards, forms a two-digit number, and covers the formed number on the 100 Chart. Then the pair adds or subtracts ten or one from the just-covered number. The pair states the resulting equation and covers the answer.

   Example: After drawing 3 and 4, the pair arranges them to make 43 or 34 and covers the selected number. If choosing to make 43, the pair could state, "43 − 10 = 33" and also cover 33.

2. The other pair draws two Digit Cards, forms a two-digit number, and covers this number (with a different-colored marker) on the same 100 Chart. Next this pair adds or subtracts ten or one, states the equation, and covers the answer.

   Example: After drawing 2 and 7, the pair may cover 27 on the board, state "27 − 1 = 26", and also cover 26.

3. Pairs take turns until one pair completes a path. A completed path may include squares that share only a common corner.

4. A pair might cover only the number made with the Digit Cards if the sums and differences are already covered. If no squares can be covered by creating a number with the two cards drawn, the pair draws a third Digit Card and uses any two of the three digits.

Making Connections:

Promote reflection and make mathematical connections by asking:

- What patterns did you notice?

Will It Fit?

Topic: Place Value

Object: Form a two-digit number within an identified range.

Groups: 2 players

Tip If children seem ready, extend this into a three-digit game.

Materials for each group

- set of Digit Cards, p. 148
- *Will It Fit?* gameboard, p. 87

Directions

1. One player mixes the Digit Cards and gives each player three cards.
2. The player whose first name comes first alphabetically uses two of the three cards to form the lowest possible two-digit number.
3. The first player announces the formed number and places the selected Digit Cards in the boxes at the top of the gameboard.
4. The other player uses two of his or her three cards to form the highest possible two-digit number.
5. The other player announces a formed number and places the selected Digit Cards in the bottom boxes.
6. The first player draws the next Digit Card. Keeping in mind that the center number needs to fit between the two displayed numbers, together the players decide whether to place the drawn card in the ones or tens place.
7. The last card is drawn and placed in the remaining box. Players win the round if this new two-digit number fits between the displayed low number and high number.
8. A player gathers all the Digit Cards, mixes them again, and plays another round.

Making Connections

Promote reflection and make mathematical connections by asking:

- Which digits were preferred draws at the beginning of the game? Why?
- How would you like to change this game?

86   *Nimble with Numbers*   Game

Will It Fit?

Low

tens   ones

High

Game   Place Value   87

Where?

Topic: Place Value

Object: Put three numbers in a row, following the correct number sequence.

Groups: Pair players or two players

Materials for each group

- *Where?* recording sheet, p. 89
- 2 colored pencils or pens (contrasting colors)
- Number Cube (1–6), p. 151
- special Number Cube (3–8), p. 151

Tip Some children appreciate the use of the 100 Chart, p. 146, as a reference when determining or challenging placement of selected numbers.

Directions

1. Ask children to recall the *Where Will It Fit?* Sponge and the importance of carefully placing and ordering numbers in each row from smallest to largest. Remind children that the winner is the first pair to get three numbers in a row, vertically, horizontally, or diagonally. Help children identify the range of numbers for each row and realize that extra care is required when placing numbers in rows two and three.

2. The first pair rolls the two number cubes, announces the possible two-digit numbers, decides which number to use, and indicates where to place that number. The other pair agrees that the placement is correct before the first pair records the number with its color pencil. If the second pair challenges the placement, that pair must explain why.

Example: If 6 and 3 are rolled, the pair may select and appropriately record 36 in column 3 of row 2, or 63 in column 4 of row 4.

3. The other pair rolls the two cubes and follows the same procedure, recording the number using a contrasting color.

4. Pairs continue alternating turns and following these steps until one pair records three of its numbers in a row. When a pair's roll produces numbers that cannot be placed, that pair passes.

Making Connections

Promote reflection and make mathematical connections by asking:

- What made it difficult to block your opponent?
- What strategies helped you line up three numbers in a row?

11				29
**30				39
**40				49
50			63	69
70				89

88   Nimble with Numbers   Game

Where?

	11						29
**	30						39
**	40						49
	50						69
	70						89

Game Place Value 89

What Numbers Are Missing? I

Use the few numbers shown in this 100 chart to write the missing numbers in the empty spaces.

								9	
	12				16				
				25				29	
31					36				
							48		
		53						59	
61						67			
				75					80
	82						88		
								99	

	2				6				
				15				19	
			34						
									50
						57		59	
61					66				
			74				78		

90 *Nimble with Numbers* Independent Activity

What Numbers Are Missing? II

Each of these is a piece cut from a 100 chart. Fill in the missing numbers.

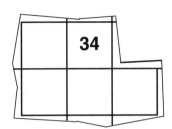
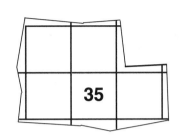

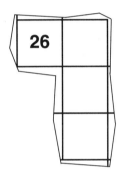
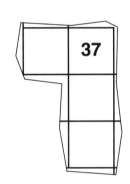
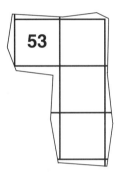

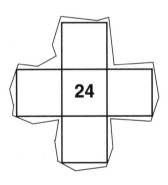
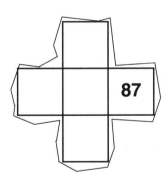
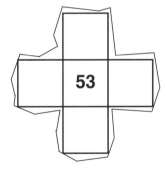

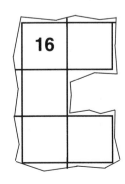
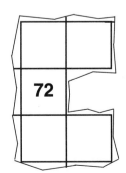
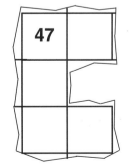

Independent Activity Place Value 91

What Numbers Are Missing? III

Each of these is a piece cut from a 100 chart. Fill in the missing numbers.

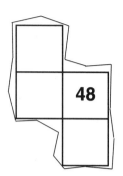
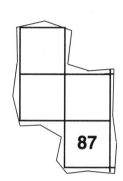

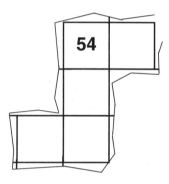
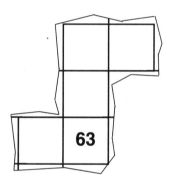
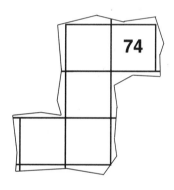

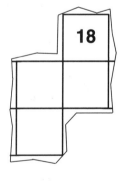
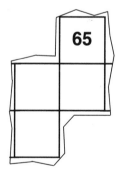
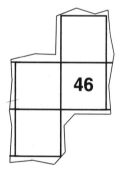

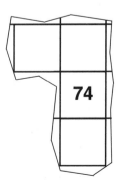
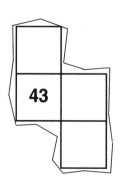
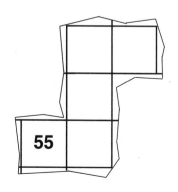

Date _____   Name _____

Staircase Number Puzzle I

Use the clues to solve this puzzle and discover some trivia.

Down Clues
2. 10 more than 26
4. 10 less than 95
6. 30 greater than 62
8. 100 less than 146

Across Clues

Use your 100 Chart to help find these numbers.

1. 44 ↓ ← = _____
3. 38 ↓ ↓ ↓ = _____
5. 70 ↑ ← = _____
7. 36 ↑ ← ← = _____
9. 59 ↓ ← ← = _____

Trivia: Number 7 across tells the number of teeth in an elephant's mouth.

Independent Activity   Place Value 93

Date _____ Name _____

Staircase Number Puzzle II

Use the clues to solve this puzzle.

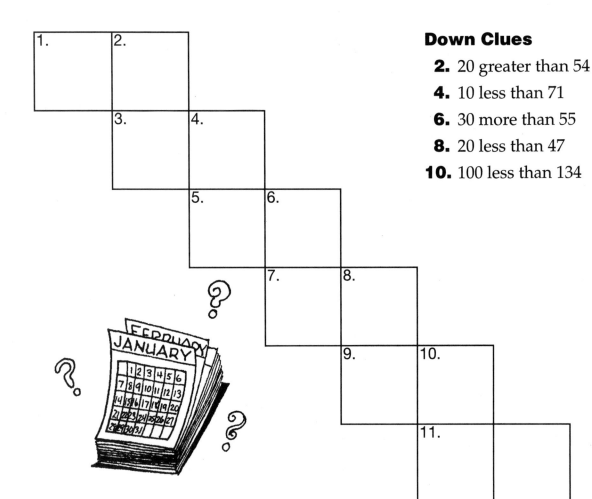

Down Clues

2. 20 greater than 54
4. 10 less than 71
6. 30 more than 55
8. 20 less than 47
10. 100 less than 134

Across Clues

Use your 100 Chart to help find these numbers.

1. 45 ↓ → →   = _____

3. 58 ↑ ← ←   = _____

5. 37 ↑ ↑ →   = _____

7. ??? (See Hint below.)

9. 84 ↑ ↑ ↓ ←   = _____

11. 70 ↑ ← ↑   = _____

Hint: The answer to Number 7 is the number of weeks in a year.

94 Nimble with Numbers   Independent Activity

Date _____ Name _____

Creating Numbers I

Use two of the three digits for each answer.

Use 2, 3, or 5.

1. Make a number less than 45. _____ _____

2. Make an odd two-digit number. _____ _____

3. Make a number greater than 35. _____ _____

4. Make the smallest possible two-digit number. _____ _____

Use 3, 4, or 9.

5. Make a number ten more than 33. _____ _____

6. Make a number between 30 and 40. _____ _____

7. Make the largest possible two-digit number. _____ _____

Use 2, 5, or 8.

8. Make a number ten less than 92. _____ _____

9. Make a number between 40 and 60. _____ _____

10. Make the smallest possible even number. _____ _____

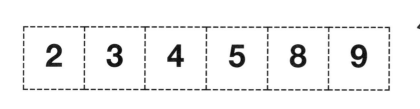

Independent Activity

Place Value 95

Creating Numbers II

Use three of the four digits for each answer.

Use 1, 4, 6, or 7.

1. Make a number greater than 650.

2. Make an even number between 400 and 600.

3. Make an odd number less than 400.

4. Make the smallest possible number.

Use 2, 5, 7, or 8.

5. Make a number less than 400.

6. Make a number between 700 and 800.

7. Make the largest possible odd number.

8. Make an even number greater than 800.

Use 1, 3, 6, or 9.

9. Make a number between 500 and 700.

10. Make an even number less than 300.

11. Make an odd number that is between 300 and 500.

12. Make the largest possible even number.

Addition

Assumptions Children have successfully used addition to solve problems given in context. Concrete objects and visual models, such as coins and ten frames, have been used extensively. Since an effort has been made to develop number sense and operation sense, children have discovered more than one meaningful way to add. It is also assumed that children have had repeated experiences to improve their abilities to mentally compute.

Section Overview and Suggestions

Sponges

Tossing Sums pp. 98–99

Add It Up pp. 100–101

These whole-class or small-group warm-ups emphasize mental addition, which is continued throughout the section. Both Sponges are open-ended and very repeatable.

Skill Checks

Sum It Up 1–6 pp. 102–104

The Skill Checks provide a way to help parents, children, and you see children's improvement with addition. Remember to have children respond to STOP before they solve the ten problems. The horizontal problems promote mental computation.

Games

18 Plus and 26 Plus pp. 105–107

Slide and Sum pp. 108–109

Target 50 pp. 110–111

All Games require strategic thinking and much addition. Because these Games are repeatable and actively involve children in mental addition, parents will enjoy playing them several times with their children.

Independent Activities

Loop Addition pp. 112–113

Addition Trees pp. 114–115

Making Sums pp. 116–117

Searching for 30 p. 118

Loop Addition, Making Sums, and *Searching for 30* require much independent addition practice as children search for addends to equal given sums. Repeated experience with the *Tossing Sums* Sponge should ensure success with all of these Independent Activities. The second version of these Independent Activities provides a familiar format with more challenging sums for children who are ready. Some of the sums for *Loop Addition* can be made in more than one way.

Tossing Sums

Topic: Mental Addition

Object: Identify three addends that equal a given sum.

Groups: Whole class or small group

Materials

- transparency of *Tossing Sums*, p. 99
- prepared listing of possible sums for three specified numbers

Tip If appropriate, extend this activity to use four rings. Some possible scores using 8, 10, and 20 with four rings: 32, 34, 36, 38, 44, 46, 48, 50, 56, 58, 60, 68, 70, 80.

Directions

1. Leader records three addends on the transparency (such as 8, 10, and 20).
2. The leader announces a total that can be made by tossing two rings on the labeled ring holders. More than one ring can be tossed on a number.
3. Ask children how a certain score, such as 30, can be scored by tossing two rings. Students mentally try to determine the location of the two rings.

   Other possible scores using 8, 10, and 20 with *two* rings: 16, 18, 20, 28, 40

4. The leader announces a total that can be made by tossing three rings on the labeled ring holders.
5. The leader asks children how a certain score such as 40, can be scored by tossing three rings. Children mentally try to determine the location of the three rings.

   Other possible scores using 8, 10, and 20 with *three* rings: 24, 26, 28, 30, 36, 38, 40, 48, 50, 60

6. Leader records three different addends on the transparency, and repeats steps 4 and 5.

   An additional starter set: 10, 15, 25

   Some possible scores using 10, 15 and 25 with *three* rings: 30, 35, 40, 45, 50, 55, 60, 65, 75

Making Connections

Promote reflection and make mathematical connections by asking:

- Which scores were easier to find? Please explain.
- What strategy helped you quickly find the addends?

Tossing Sums

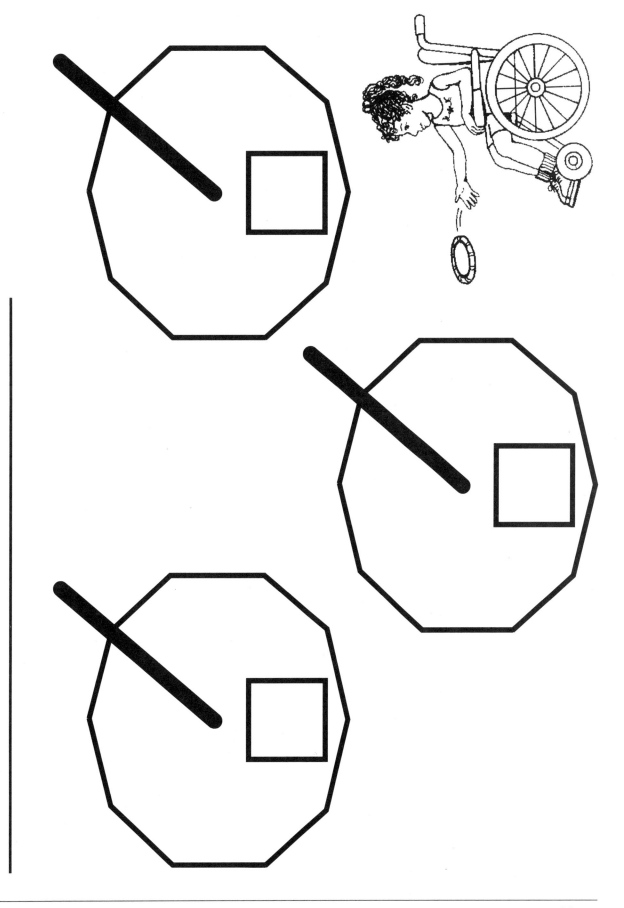

Sponge Addition 99

Add It Up

Topic: Mental Addition

Object: Create addition problems that produce sums to meet given criteria.

Groups: Whole class or small group

Materials

- transparent set of Digit Squares, p. 149
- Digit Squares for each child, p. 149
- transparency of *Add It Up* with forms cut apart, p. 101

Directions

1. Leader describes the criteria for the desired sum.

   Example: "Create an addition problem that produces a sum greater than 50."

2. Leader displays the addition-problem format that is copied by the children.

3. Leader draws the first two Digit Squares and announces the digits. Each child places the announced digits on his or her *Add It Up* form. If preferred, one of the announced digits can be placed in the "reject circle." (Drawn digits are not reused.)

4. Leader makes sure each child places the digits in two of the squares and then draws and announces two additional Digit Squares.

5. After children have placed four Digit Squares, children total their addition problems to determine which sums meet the criteria.

6. Children gather and mix the transparent Digit Squares to prepare for another round of this activity.

   Possible criteria for future rounds: Sum between 60 and 80; Odd-numbered sum greater than 20; Sum close to 50; Largest sum; or Lowest sum.

7. If the more challenging addition-problem format is used from page 101, leader must draw five digits.

Making Connections

Promote reflection and make mathematical connections by asking:

- How did you decide where to place certain digits?

Add It Up

Add It Up

Sponge   Addition   101

Date _____   Name _____

Sum It Up 1

STOP Don't start yet! Star two problems that may have even answers.

1. 15 + 6 = ____   **2.** 21 + 5 + 4 = ____   **3.** 33 + 8 = ____

4. 47 + 6 = ____   **5.** 55 + 24 = ____

6. 34   **7.** 49   **8.** 26   **9.** 45   **10.** 137
   + 8   + 7   + 24   + 37   + 5
   ___   ___   ___   ___   ___

Go On What numbers come next? 15, 18, 21, 24, ____ , ____

Date _____   Name _____

Sum It Up 2

STOP Don't start yet! Star two problems that may have answers less than 50.

1. 19 + 6 = ____   **2.** 42 + 6 + 4 = ____   **3.** 64 + 8 = ____

4. 27 + 5 = ____   **5.** 36 + 22 = ____

6. 45   **7.** 56   **8.** 37   **9.** 47   **10.** 145
   + 6   + 9   + 34   + 26   + 9
   ___   ___   ___   ___   ___

Go On What number is missing? 21, 25, 29, ___, 37, 41

102   Nimble with Numbers   Skill Checks

Date _____   Name _____

Sum It Up 3

STOP Don't start yet! Star the problem that may have the greatest answer.

1. 15 + 7 = _____   **2.** 32 + 4 + 5 = _____   **3.** 54 + 9 = _____

4. 43 + 9 = _____   **5.** 56 + 23 = _____

6. 63   **7.** 68   **8.** 45   **9.** 37   **10.** 125
    + 8   + 7   + 27   + 48   + 6

Go On Write three equations that equal 25.

Date _____   Name _____

Sum It Up 4

STOP Don't start yet! Star two problems that may have odd answers.

1. 18 + 3 =   **2.** 26 + 4 + 3 =   **3.** 55 + 7 =

4. 52 + 9 =   **5.** 35 + 24 =

6. 54   **7.** 79   **8.** 57   **9.** 38   **10.** 143
    + 8   + 9   + 25   + 28   + 8

Go On Write another equation that fits. Please explain your answer.

| 35 + 15 = | 33 + 17 = |
| 28 + 22 = | |

Skill Checks   Addition 103

Date _____ Name _____

Sum It Up 5

 Don't start yet! Star three problems that may have even answers.

1. 17 + 7 = _____   **2.** 41 + 6 + 3 = _____   **3.** 28 + 6 = _____

4. 34 + 9 = _____   **5.** 54 + 35 = _____

6. 49
  + 4
  ___

7. 65
  + 9
  ___

8. 48
  + 25
  ___

9. 48
  + 39
  ___

10. 165
  + 7
  ___

Go On ➡ Write three equations that equal 30.

Date _____ Name _____

Sum It Up 6

 Don't start yet! Star the problems that may have the lowest answer.

1. 19 + 3 = _____   **2.** 34 + 5 + 4 = _____   **3.** 35 + 6 = _____

4. 73 + 8 = _____   **5.** 64 + 13 = _____

6. 65
  + 7
  ___

7. 89
  + 8
  ___

8. 36
  + 34
  ___

9. 37
  + 47
  ___

10. 158
  + 5
  ___

Go On ➡ What number comes next? 23, 28, 33, 38, _____ , _____

18 Plus

Topic: Addition

Object: Cover three in a row with your markers.

Groups: Pair players or 2 players

Materials for each group

- *18 Plus* gameboard, p. 106
- markers (different kind for each pair)
- special Number Cube (4–9), p. 152

Tip To extend the playing and practice time, require winners to cover 4 markers in a row.

Directions

1. The first pair rolls the special Number Cube to determine the amount to add to 18. After the pair states the equation including the sum, they place one of their markers on the number that equals the sum.

   Example: If 7 is rolled, pair states, "18 plus 7 equals 25," and covers one of the 25s on the gameboard.

2. Pairs alternate turns, following this procedure until one pair wins by placing three of their markers in a row horizontally, vertically, or diagonally.

3. To provide additional practice with a different 2-digit addend, use the *26 Plus* gameboard. *26 Plus* also requires the use of a 4–9 Number Cube (or spinner).

26	24	23	27
●25	●22	26	23
23	27	25	○24
25	24	27	22

Making Connections

Promote reflection and make mathematical connections by asking:

- What strategies did you use in placing your markers?
- Were some numbers easier to cover than others?
- Was it difficult to block your opponent? Why or why not?

18 Plus

26	24	23	27
25	22	26	23
23	27	25	24
25	24	27	22

26 Plus

32	34	31	35
35	33	34	30
34	35	32	33
30	32	30	31

Game

Addition 107

Slide and Sum

Topic: Mental Addition

Object: Add to a specified target sum.

Groups: Pair players or 2 players

Materials for each group

- *Slide and Sum* gameboard, p. 109
- one marker

Tip Encourage strategic thinking by having players keep the same target sum for multiple rounds.

Directions

1. Players pick a target sum between 20 and 30.
2. The first pair places the marker on any number and announces that number.
3. The second pair slides the same marker along a line in any direction to identify the next number. The pair adds this number to the previous number and states the total.
4. The pairs alternate turns by sliding the marker to another number, adding that amount to the previous total, and announcing the new total. Pairs must move the marker and give a new total on each turn. Pairs may return to previously used numbers.
5. The pair who states the target sum as their new total wins that round.
6. If a pair is forced to exceed the target sum, the game ends without a winner.

Making Connections

Promote reflection and make mathematical connections by asking:

- Where is a good starting place?
- What strategy helped you reach the target sum?
- If zero were added to the gameboard, where should it be placed?

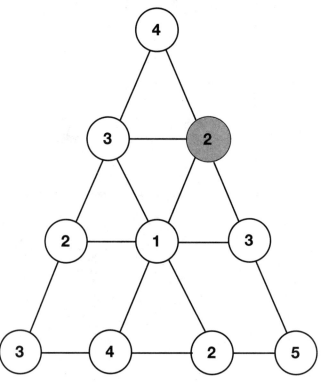

108 Nimble with Numbers   Game

Slide and Sum

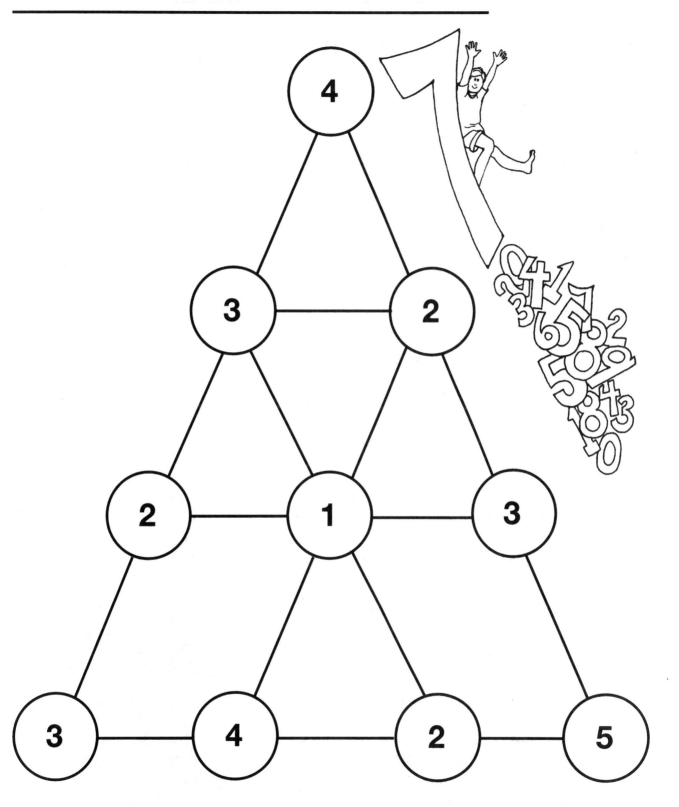

Game Addition 109

Target 50

Topic: Mental Addition to 50

Object: Reach close to the sum of 50.

Groups: Pair players or 2 players

Materials for each pair

- *Target 50* recording sheet, p. 111
- set of Digit Squares (only the digits 1–5), p. 149
- opaque container for Digit Squares

Directions

1. Each pair has a recording sheet and a container with Digit Squares 1–5. The first pair draws one Digit Square and decides whether to put it in the tens place or ones place. Once a square is placed it cannot be moved. To visualize how close a pair is to 50, the pair shades the grid on the recording sheet after each turn.
2. The other pair follows the same procedure, drawing a Digit Square and placing it on the recording sheet.
3. Pairs alternate turns by drawing and placing Digit Squares two more times.
4. After placing three Digit Squares, each pair totals the results and records the difference between the total and 50.
5. The differences found in step 4 are the scores. The pair with the lower score wins.

Making Connections

Promote reflection and make mathematical connections by asking:
- Why do you think the player with the lower score wins?
- How did you decide where to place your digits?
- How would you play differently in future games?

Tip If children become confident playing this version, challenge them with higher target numbers. For example, have them draw five Digit Squares and extend the target number to 75.

Target 50

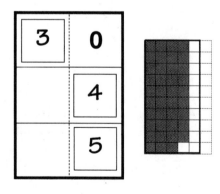

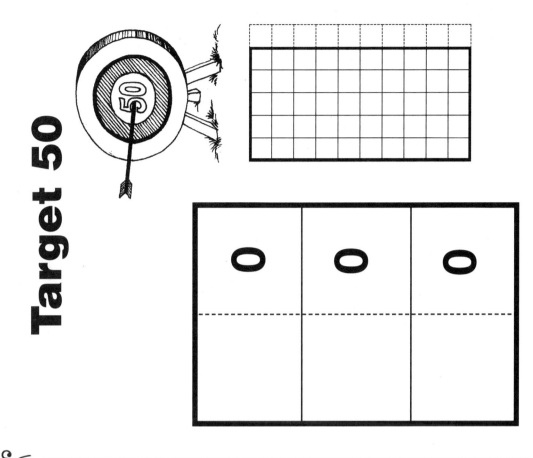

Loop Addition I

Draw loops around two groups of numbers to match the sum. See the example.

Example
15 3 3 ③
    ④ 4 ④
3 + 4 + 4 + 4 = 15
	10 2 2 2
	    3 3 3
		11 2 2 2
		    3 3 3

12 2 2 2
    4 4 4
	14 2 2 2
	    4 4 4
		16 2 2 2
		    4 4 4

14 3 3 3
    4 4 4
	17 3 3 3
	    4 4 4
		18 3 3 3
		    4 4 4

Draw a loop around one group of numbers to match the sum.

Example: Sum = 12

②
③
⑦
6

Sum = 11	Sum = 12	Sum = 13	Sum = 10	Sum = 14	Sum = 13	Sum = 15
6	3	7	2	2	3	4
2	6	1	5	7	4	3
3	4	5	0	3	5	8
4	2	4	2	4	4	5
		2	3	4	1	3

112 Nimble with Numbers — Independent Activity

Loop Addition II

Draw loops around two groups of numbers to match the sum. See the example.

Example
15 3 3 (3)
   (4 4) (4)
3 + 4 + 4 + 4 = 15
	18 4 4 4
	    5 5 5
		20 4 4 4
		    6 6 6

20 3 3 3
    7 7 7
	23 3 3 3
	    7 7 7
		24 3 3 3
		    7 7 7

22 5 5 5
    7 7 7
	24 5 5 5
	    7 7 7
		29 5 5 5
		    7 7 7

Draw loops around one group of numbers to match the sum.

Example: **Sum = 16**

3
(5
7
4)
2

Sum = 19	Sum = 21	Sum = 20	Sum = 23	Sum = 22	Sum = 26	Sum = 25
4	4	4	7	1	7	9
5	7	7	8	7	6	8
5	6	5	5	8	6	7
5	8	1	8	7	7	5
3	5	4	2	5	5	5
		3	6	1	4	6

Independent Activity   Addition 113

Date _____   Name _____

Addition Trees I

The first problem is done for you.

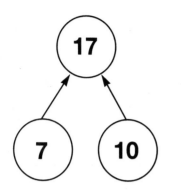
```
   7
+ 10
----
  17
```

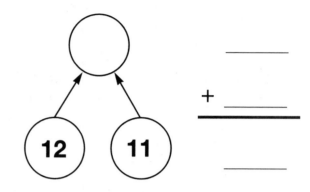

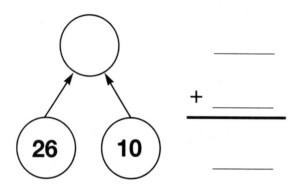

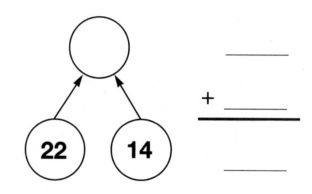

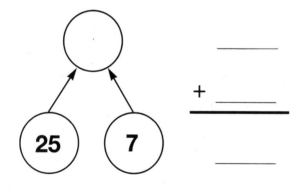

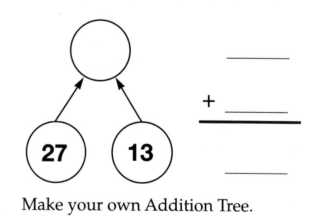

Make your own Addition Tree.

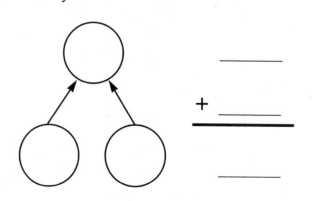

Date _____ Name _____

Addition Trees II

The first problem is done for you.

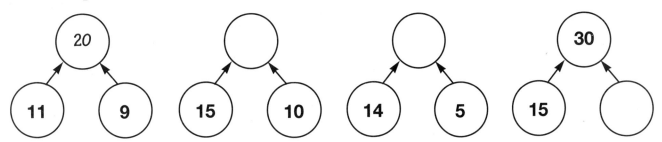

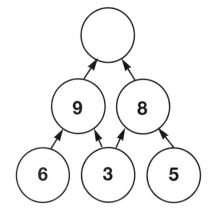
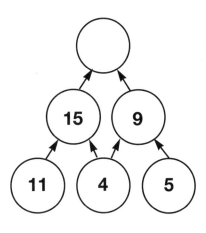
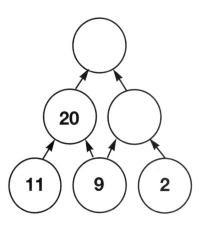

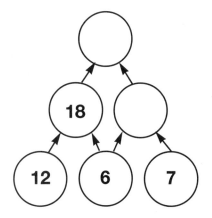
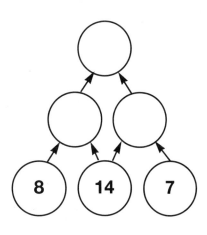
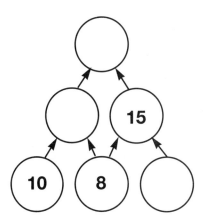

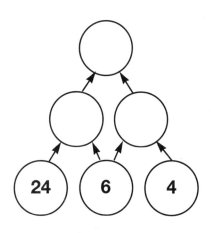

Make your own Addition Tree.

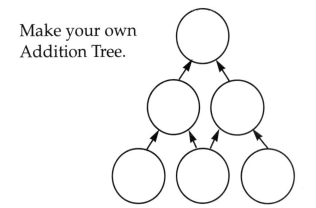

Date _____ Name _____

Making Sums I

Use the numbers in the box to create addition problems.

20	30
10	17

____ + ____ = 27   ____ + ____ = 40

____ + ____ = 50   ____ + ____ = 30

25	10
15	20

____ + ____ = 45   ____ + ____ = 40

____ + ____ = 35   ____ + ____ = 25

35	10
20	9

____ + ____ = 45   ____ + ____ = 55

____ + ____ = 44   ____ + ____ = 19

30	20
15	7

____ + ____ = 45   ____ + ____ = 35

____ + ____ = 22   ____ + ____ = 50

Make a *new sum* using two of the numbers 30, 20, 15, and 7.

_____ + _____ = _____

Making Sums II

Use the numbers in the box to create addition problems.

10	15
17	20

____ + ____ = 30   ____ + ____ = 32
____ + ____ = 35   ____ + _10_ + ____ = 45

15	10
20	9

____ + ____ = 30   ____ + ____ = 25
____ + ____ = 24   _15_ + ____ + ____ = 34

30	11
25	15

____ + ____ = 41   ____ + ____ = 40
____ + ____ = 45   ____ + ____ = 26
____ + ____ + _11_ = 51   ____ + _30_ + ____ = 70

20	30
25	9

____ + ____ = 45   ____ + ____ = 39
____ + ____ + _30_ = 59   _30_ + ____ + ____ = 75

Make a *new sum* using three of the numbers 30, 25, 20, and 9.

_____ + _____ + _____ = _____

Independent Activity   Addition 117

Date _____ Name _____

Searching for 30

Find three in a row that equal 30. One is done as an example.

(14)	6	10	7	15
5	13	6	11	6
7	9	14	7	9
15	8	8	12	10

Search Results

5-6 Good

7-8 Great!

9-10 Terrific!!

11 Wow!!!

Record your findings.

Across ▯▯▯

_14_ + _6_ + _10_ = 30   ____ + ____ + ____ = 30

____ + ____ + ____ = 30   ____ + ____ + ____ = 30

Down

____ + ____ + ____ = 30   ____ + ____ + ____ = 30

____ + ____ + ____ = 30   ____ + ____ + ____ = 30

Diagonals

____ + ____ + ____ = 30   ____ + ____ + ____ = 30

____ + ____ + ____ = 30   ____ + ____ + ____ = 30

118 Nimble with Numbers   Independent Activity

Subtraction

Assumptions Children have successfully used subtraction to solve problems given in context. Since an effort has been made to develop number sense and operation sense, children have discovered more than one meaningful way to subtract. It is also assumed that children have had repeated experiences to improve their abilities to mentally compute.

Section Overview and Suggestions

Sponges

Three Ten Frames pp. 120–122

Make the Difference pp. 123–125

Finding Differences p. 126

Three Ten Frames provides a visual model while the other two warm-ups reinforce mental subtraction. *Three Ten Frames* and *Make the Difference* are open-ended and very repeatable, while *Finding Differences* allows many possibilities.

Skill Checks

What's the Difference? 1–6 pp. 127–129

These Skill Checks provide a way to help parents, children, and you see children's improvement with subtraction. Remember to have all children respond to STOP before they solve the ten problems. The horizontal problems promote mental computation.

Games

Subtract and Travel pp. 130–132

Differences Count pp. 133–135

Difference Bingo pp. 136–138

These are repeatable, engaging Games that parents will enjoy playing several times with their children. All three Games require strategic thinking and actively involve children in much mental subtraction. Each Game provides a more challenging version.

Independent Activities

Subtraction Squares pp. 139–140

Finding Pairs pp. 141–142

Rearrange and Find pp. 143–144

Each Independent practice Activity includes a self-check feature for the children. The second version of each Independent Activity provides a greater challenge. *Finding Pairs* extends skills children developed while experiencing the *Finding Differences* sponge. Some problems can be solved in more than one way. Children should be encouraged to use the movable Digit Squares to solve the *Rearrange and Find* problems.

Three Ten Frames

Topic: Subtraction

Object: Visualize and verbalize differences.

Groups: Whole class or small group

Materials

- transparency of *Three Ten Frames*, p. 121
- transparency of *Ten Frame Patterns* (cut apart), p. 122

Directions

1. Leader displays one full Ten Frame and one partial Ten Frame (with 8 cells filled) on the *Three Ten Frames* transparency. Leader asks, "How much is shown?"
2. After children identify the amount shown, the leader asks, "How many more would you need to make 26?" As children respond, leader asks them to explain what they did mentally or visually to find the difference.
3. Leader asks children for the related subtraction equation that illustrates the problem, *26 – 18*. (When subtracting, the total is known, as well as one of the two parts.)
4. Leader poses similar problems, starting with a display of 18 and finding how many more are needed to make a variety of totals. For each of these totals, children help the leader identify the related subtraction problem.
5. Leader poses similar problems with a new starting amount.

   Example: Display 16 and ask, "How many more are needed to have a total of 23?"
6. Remember to have children identify the related subtraction problem for each problem posed.

Tip As children become more competent, begin with larger amounts, such as 27, and make totals beyond 30. This requires children to mentally visualize additional Ten Frames.

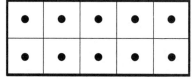

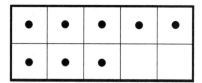

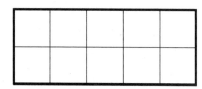

Making Connections

Promote reflection and make mathematical connections by asking:

- How do Ten Frames help you find differences?

Three Ten Frames

Sponge Subtraction 121

Ten Frame Patterns

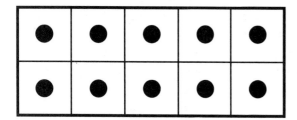
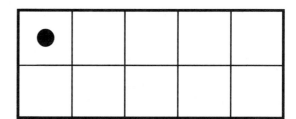
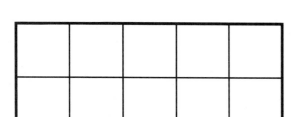
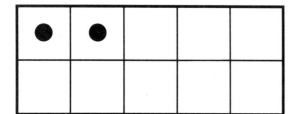
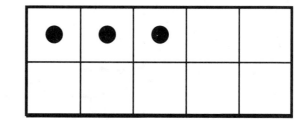
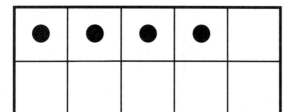
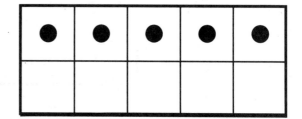
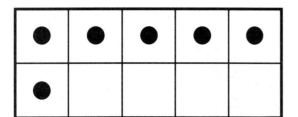
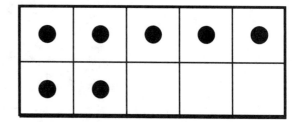
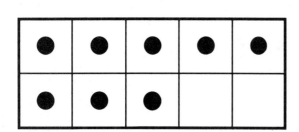

122 *Nimble with Numbers* **Sponge**

Make the Difference

Topic: Mental Subtraction

Object: Create subtraction problems that produce differences to meet given criteria.

Groups: Whole class or small group

Materials

- transparent set of Digit Cards, p. 148
- container for Digit Cards
- transparency of *Make the Difference* forms, pp. 124–125

Tip If desired, leader can draw an additional Digit Card each round and children can discard one drawn number in a "reject circle."

Directions

1. Leader describes the criteria for the desired difference.

   Example: Create a subtraction problem that produces a difference close to 30.

2. Leader displays a format for the subtraction problem which is copied by the children (see illustration). Leader explains that four cards will be separately drawn and announced.

3. Leader draws the first card and announces the digit. Before a new number is drawn, leader makes sure that each child records the previous digit in one of the squares. Once a number is recorded it cannot be changed.

4. The leader continues to follow this process until four cards are drawn. (Drawn cards are not reused until the next round.)

5. After players have recorded the four digits, children compute the difference of their subtraction problems to determine which differences are closest to 30.

6. Players continue to play additional rounds of this activity. *Possible criteria for future rounds:* Difference between 10 and 20; Odd-numbered difference greater than 20; Difference closest to 50; Smallest difference; or Largest difference

7. If an easier subtraction format is desired (p. 125), leader must then draw three Digit Cards.

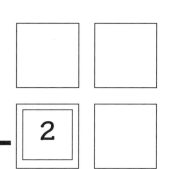

Making Connections

Promote reflection and make mathematical connections by asking:

- How did you decide where to place certain digits?

Sponge

Make the Difference

124 *Nimble with Numbers* **Sponge**

Make the Difference

Sponge Subtraction 125

Finding Differences

Topic: Mental Subtraction

Object: Identify two numbers whose difference equals a given amount.

Groups: Whole class or small group

Materials

- chalkboard or overhead projector

Tip For additional experience with this activity, have children do the Finding Pairs activity, pp. 141–142.

Directions

1. Leader displays a group of numbers. (See first boxed list.)
2. Leader announces a possible difference, such as 20.
3. Children try to identify two listed numbers whose difference is 20. Leader asks, "What other ways can it be done?"

   Example: 47 − 27 = 20 or 67 − 47 = 20 or ...

4. Leader continues stating differences and children respond with possibilities. (Other possible differences for the first box: 10, 30, 40, 50.)
5. Challenge children with a slightly varied listing. (See second boxed list.) Possible differences for the second box: 5, 10, 15, 20, 25, 30, 35, 40, 45. (Some differences can be found more than one way.) An additional challenging arrangement:

   Boxed numbers: 46, 9, 26, 56, 36, 16

   Possible differences: 7, 10, 17, 20, 27, 30, 37, 40, 47

27	67
57	37
17	47

18	13
48	38
58	28

Making Connections

Promote reflection and make mathematical connections by asking:

- What patterns did you notice?
- What helped you quickly locate differences?

"How could you make a difference of 20?"

Date _____ Name _____

What's the Difference? 1

 Don't start yet!
Star three problems that may have even answers.

1. 25 − 7 = ____ **2.** 31 − 20 = ____ **3.** 51 − 6 = ____

4. ☐ + 13 = 20   **5.** 32 − 19 = ____

6. 45 **7.** 50 **8.** 67 **9.** 146 **10.** 240
− 7   − 21   − 28   − 8   − 26

 What number comes next? 74, 72, 70, ____
Please identify the pattern.

Date _____ Name _____

What's the Difference? 2

 Don't start yet!
Star two problems that may have answers between 20 and 40.

1. 34 − 9 = ____ **2.** 48 − 5 = ____ **3.** 43 − 20 = ____

4. 14 + ☐ = 20   **5.** 44 − 19 = ____

6. 47 **7.** 40 **8.** 57 **9.** 251 **10.** 150
− 9   − 19   − 18   − 6   − 23

 Which number is closer to 100: 104 or 97?
Please prove it.

Skill Checks Subtraction 127

Date _____ Name _____

What's the Difference? 3

 Don't start yet!
Star two problems in Row 1 that may have even answers.

1. 68 – 30 = _____   **2.** 44 – 6 = _____   **3.** 32 – 8 = _____

4. □ + 15 = 20   **5.** 33 – 29 = _____

6. 63   **7.** 50   **8.** 41   **9.** 336   **10.** 160
    – 8   – 14   – 25   – 8   – 37
    ___   ____   ____   ____   ____

 What number is missing? 66, 61, 56, ____, 46, 41

Date _____ Name _____

What's the Difference? 4

 Don't start yet!
Star two problems in Row 3 that may have odd answers.

1. 42 – 30 = _____   **2.** 43 – 9 = _____   **3.** 67 – 4 = _____

4. 8 + □ = 20   **5.** 43 – 19 = _____

6. 52   **7.** 30   **8.** 62   **9.** 191   **10.** 140
    – 7   – 18   – 34   – 7   – 15
    ___   ____   ____   ____   ____

Write three subtraction equations that equal 23.

Date _____ Name _____

What's the Difference? 5

 Don't start yet!
Star the problems in Row 3 that may have answers greater than 50.

1. 21 – 9 = _____   **2.** 38 – 10 = _____   **3.** 34 – 5 = _____

4. ☐ + 12 = 20   **5.** 42 – 19 = _____

6. 51   **7.** 40   **8.** 53   **9.** 246   **10.** 170
  – 6   – 26   – 17   – 7   – 23
  ___   ___   ___   ___   ___

 Which number is closer to 73: 65 or 78?
Please prove it.

Date _____ Name _____

What's the Difference? 6

 Don't start yet!
Star the problem that may have the largest answer.

1. 34 – 8 = _____   **2.** 43 – 9 = _____   **3.** 49 – 30 = _____

4. 11 + ☐ = 20   **5.** 31 – 29 = _____

6. 44   **7.** 50   **8.** 34   **9.** 451   **10.** 180
  – 7   – 28   – 26   – 7   – 47
  ___   ___   ___   ___   ___

 Write three subtraction equations that equal 28.

Skill Checks　　　　　　　　　　　　　　　　　　　　　　　Subtraction 129

Subtract and Travel

Topic: Subtraction

Object: Create a pathway across the board.

Groups: Pair players or 2 players

Materials for each group

- Markers (different kind for each pair)
- special Number Cube (3-3-4-4-5-5), p. 152
- *Subtract and Travel A* gameboard, p. 131

Directions

1. After tossing the 3-3-4-4-5-5 Number Cube, the first pair looks for cells with differences that match the tossed digit. The pair announces the equation (such as "90 minus 85 equals 5") and covers one of the identified cells with a marker.

2. The other pair follows the same procedure but uses a different kind of marker. (Only one marker can occupy a single cell.)

3. Pairs continue alternating turns until one pair forms a continuous pathway *across* the gameboard (top to bottom or left to right).

Making Connections

Promote reflection and make mathematical connections by asking:

- What strategy helped you place your markers in a complete pathway?

Tip When children are ready for a greater challenge, use Subtract and Travel B *gameboard, p. 132. Note a different special number cube is required, p. 152 (5-5-6-6-7-7). The blank* Subtract and Travel *gameboard included with the Blackline Masters (p. 154) allows further variation and use of this popular Game.*

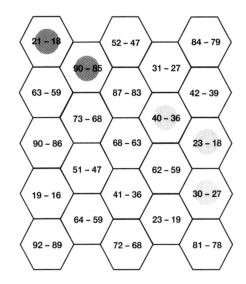

130   Nimble with Numbers   Game

Subtract and Travel A

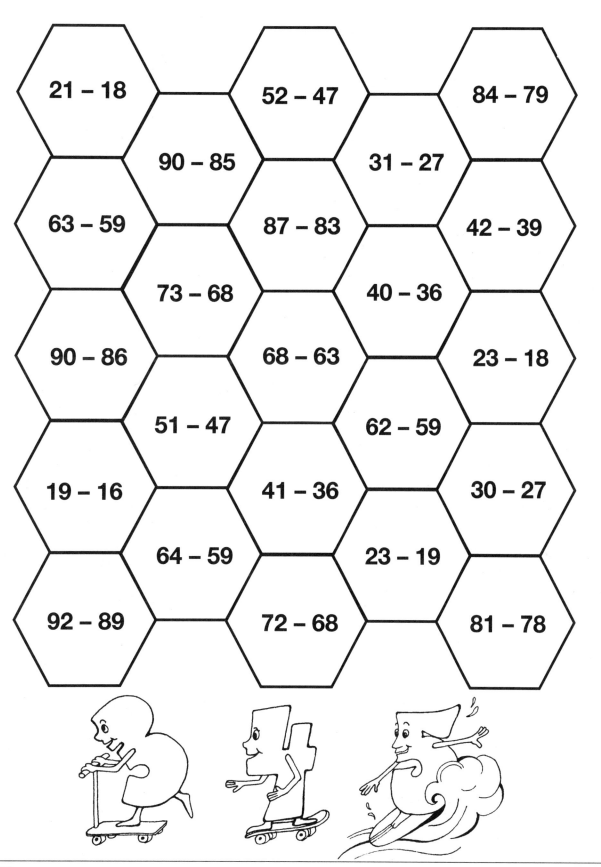

Game   Subtraction 131

Subtract and Travel B

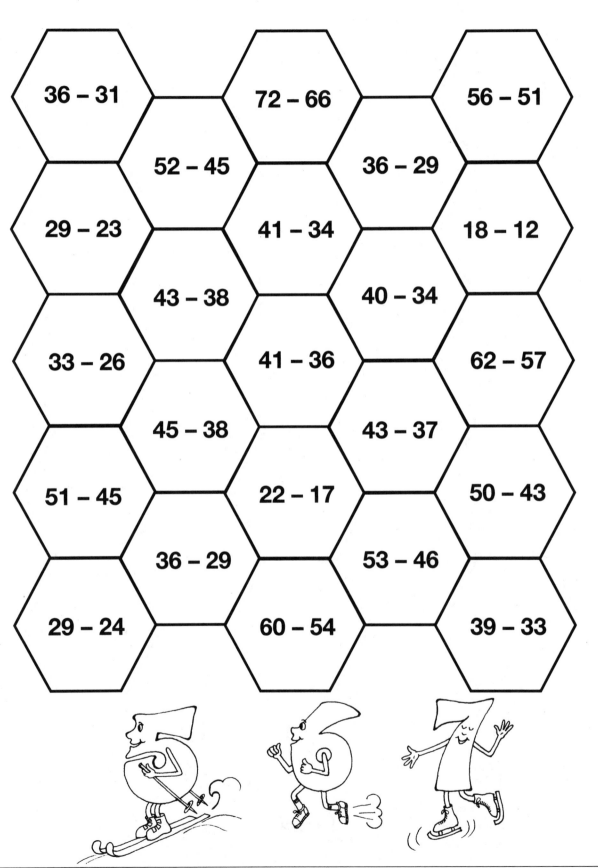

Differences Count

Topic: Mental Subtraction

Object: Create qualifying subtraction equations.

Groups: Pair players or 2 to 4 players

Materials for each group

- *Differences Count A* recording sheet (one for each pair), p. 134
- two sets of Digit Cards, p. 148

Tip For a greater challenge, use the Differences Count B recording sheet with the same materials and rules.

Directions

1. Each pair receives a recording sheet and draws four Digit Cards. Then each pair forms 2 two-digit numbers and finds the difference, thus creating an equation.

2. Looking at the bottom of the recording sheets, each pair finds the cells containing the "differences" they found. Then each pair records its equation in the appropriate cell.

   Example: One pair draws 7, 5, 2, 6; forms the problem *72 – 56;* states the equation "72 minus 56 equals 16;" and records the equation in the cell labeled *15–19* on the recording sheet.

3. Pairs continue play until one pair completes all cells on the recording sheet. After every two rounds, cards should be mixed and restacked.

4. When a pair is unable to form a needed subtraction equation, nothing is recorded for that turn.

Making Connections

Promote reflection and make mathematical connections by asking:

- What strategy did you use to help you find the difference?
- Describe the characteristics of a good draw (or a poor draw).
- Which differences were harder to find?

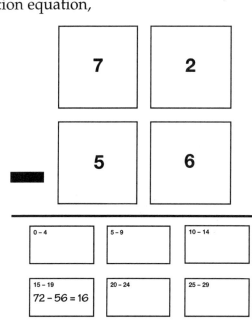

Game Subtraction 133

Differences Count A

| 0 – 4 | 5 – 9 | 10 – 14 |
| 15 – 19 | 20 – 24 | 25 – 29 |

Differences Count B

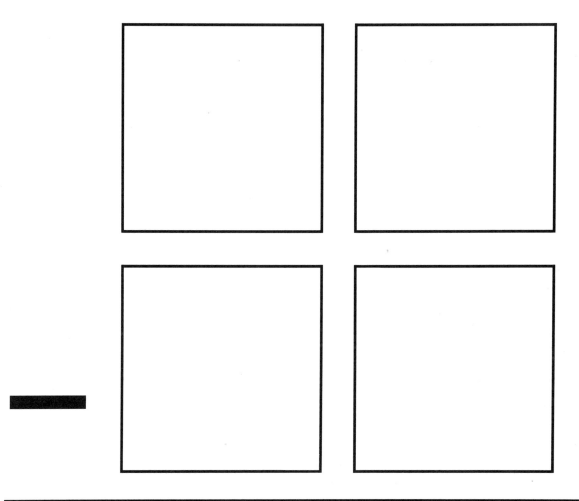

| 30 – 34 | 35 – 39 | 40 – 44 |
| 45 – 49 | 50 – 54 | 55 – 59 |

Difference Bingo

Topic: Subtraction

Object: Cover three numbers in a row.

Groups: Pair players or 2 players

Materials for each group

- *Difference Bingo A* gameboard, p. 137
- 2 sets of Digit Cards (digits 1–5 only), p. 148
- markers (a different kind for each pair)

Directions

1. Each pair draws 4 Digit Cards.
2. Each pair forms 2 two-digit numbers and finds the difference between the 2 two-digit numbers. If the difference appears on the gameboard, the pair places a marker on the difference. If not, the pair rearranges drawn Digit Cards until able to form a subtraction problem with a difference that appears on the gameboard.
3. Play continues with each pair taking turns until one pair has three markers in a row horizontally, vertically, or diagonally.
4. If it is not possible to form a subtraction problem equal to an uncovered difference, the pair does not place a marker for that turn.

Making Connections

Promote reflection and make mathematical connections by asking:

- Were some differences easier or more difficult to make than others? Why do you think this is so?
- How were you able to use patterns to help you win this game?
- What strategy did you use to help you place three markers in a row?

Tip For a greater challenge, use Difference Bingo B gameboard with one complete set of Digit Cards plus cards 0–4. Players use only 4 out of 5 drawn Digit Cards. The winner is the first player to get three markers in a row.

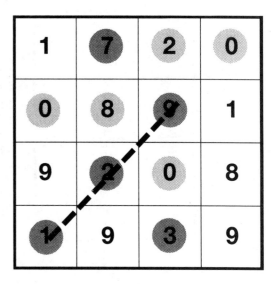

Difference Bingo A

1	7	2	0
0	8	9	1
9	2	0	8
1	9	3	9

Game   Subtraction   137

Difference Bingo B

4	9	2	11	8
11	8	5	14	0
7	12	9	4	13
10	6	3	5	7
2	1	15	10	6

Subtraction Squares III

Subtract each row and column to fill in the missing numbers.

Sample

36	20	16
19	7	12
17	13	4

54	30
49	25

37	20
18	12

42	39
31	28

38	29
15	12

46	30
27	15

56	21
19	7

45	19
35	15

28	9
19	6

Date _____ Name _____

Subtraction Squares IV

Subtract each row and column to fill in the missing numbers.

41−19=37−15

Sample

60	19	41
23	14	9
37	5	32

65	39
29	15

71	23
27	18

74	21
49	17

64	31
36	25

57	2
38	19

81	41
25	17

70	35
34	18

72	45
24	19

140 Nimble with Numbers Independent Activity

Date _____   Name _____

Finding Pairs I

Use the numbers in the box to create subtraction problems.

66	36
56	26

____ − ____ = 10   ____ − ____ = 30

____ − ____ = 20   ____ − ____ = 40

40	25
30	15

____ − ____ = 5   ____ − ____ = 15

____ − ____ = 10   ____ − ____ = 25

44	14
24	9

____ − ____ = 30   ____ − ____ = 15

____ − ____ = 5   ____ − ____ = 35

Make two **new differences** by using the numbers in the box.

45	25
35	15

____ − ____ = ____

____ − ____ = ____

Independent Activity   Subtraction   141

Date _____ Name _____

Finding Pairs II

Use the numbers in the box to create subtraction problems.

75	25
50	15

___ − ___ = 50   ___ − ___ = 10
___ − ___ = 60   ___ − ___ = 25

51	21
31	7

___ − ___ = 30   ___ − ___ = 14
___ − ___ = 44   ___ − ___ = 24

49	29
39	11

___ − ___ = 20   ___ − ___ = 38
___ − ___ = 28   ___ − ___ = 18

65	35
55	16

___ − ___ = 19   ___ − ___ = 30
___ − ___ = 39   ___ − ___ = 49

Make two **new differences** by using the numbers in the box.

42	12
22	15

___ − ___ = ___
___ − ___ = ___

142   Nimble with Numbers   **Independent Activity**

Rearrange and Find I

Place 2, 4, and 6 in the proper squares to make each difference.

```
  [4] [6]        [ ] [ ]        [ ] [ ]
 −    [2]       −    [ ]       −    [ ]
 ─────────      ─────────      ─────────
   4   4          2   2          1   8


  [ ] [ ]        [ ] [ ]        [ ] [ ]
 −    [ ]       −    [ ]       −    [ ]
 ─────────      ─────────      ─────────
   5   8          6   2          3   6
```

Place 3, 4, and 5 in the proper squares to make each difference.

```
  [ ] [ ]        [ ] [ ]        [ ] [ ]
 −    [ ]       −    [ ]       −    [ ]
 ─────────      ─────────      ─────────
   4   2          4   9          3   8
```

┌───┬───┬───┬───┬───┐
│ 2 │ 3 │ 4 │ 5 │ 6 │
└───┴───┴───┴───┴───┘

Independent Activity   Subtraction 143

Date _____ Name _____

Rearrange and Find II

Place 2, 3, and 5 in the squares to make each sum or difference.
Use + or − in the triangles.

      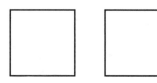

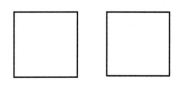   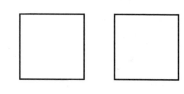

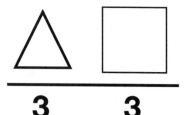

144 *Nimble with Numbers* **Independent Activity**

Blackline Masters

100 Chart
Coins
Digit Cards
Digit Squares
Dot Patterns
Number Cubes (1–6, 3–8)
Number Cubes (4–9, blank)
Spinners (1–6, blank)
Subtract and Travel Form
Subtraction Squares Form
Ten Frames
Addition Chart

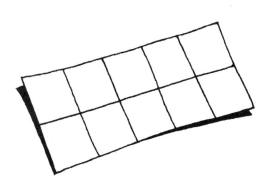

100 Chart

1	2	3	4	5	6	7	8	9	10
11	12	13	14	15	16	17	18	19	20
21	22	23	24	25	26	27	28	29	30
31	32	33	34	35	36	37	38	39	40
41	42	43	44	45	46	47	48	49	50
51	52	53	54	55	56	57	58	59	60
61	62	63	64	65	66	67	68	69	70
71	72	73	74	75	76	77	78	79	80
81	82	83	84	85	86	87	88	89	90
91	92	93	94	95	96	97	98	99	100

Coins

Digit Cards

0		
1	2	3
4	5	6
7	8	9

Digit Squares

0	1	2	3	4
5	6	7	8	9

0	1	2	3	4
5	6	7	8	9

0	1	2	3	4
5	6	7	8	9

Dot Patterns

Number Cubes

Cut solid lines. Fold on dotted lines.

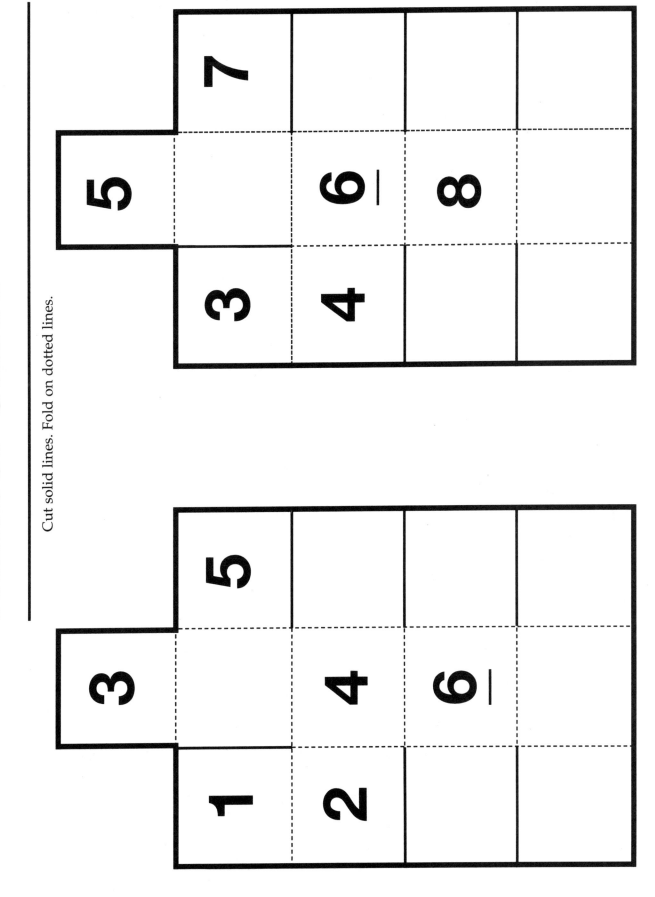

Blackline Masters 151

Number Cubes

Cut solid lines. Fold on dotted lines.

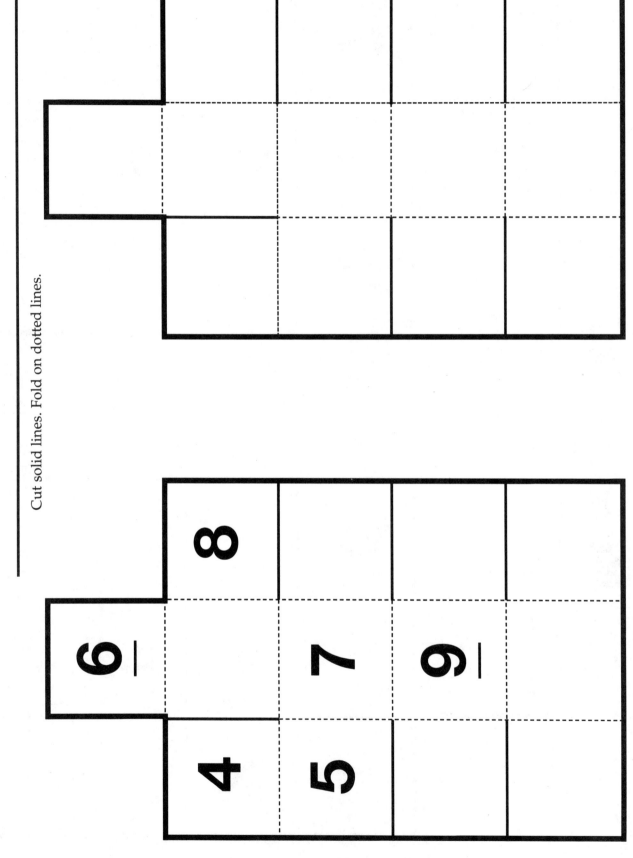

Spinners

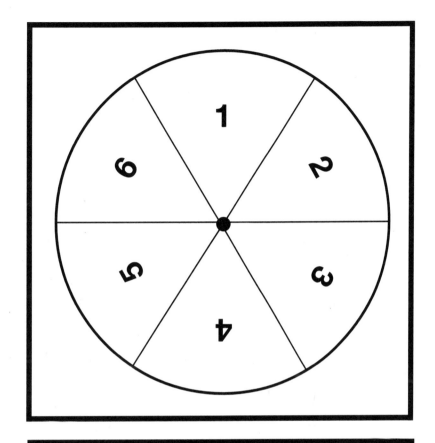

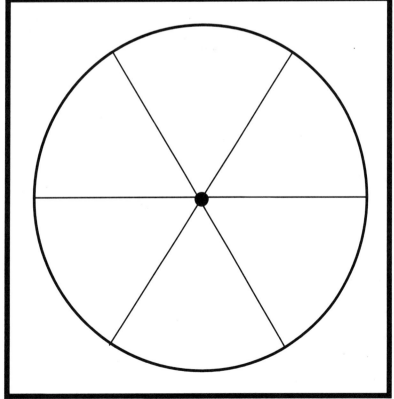

Subtract and Travel

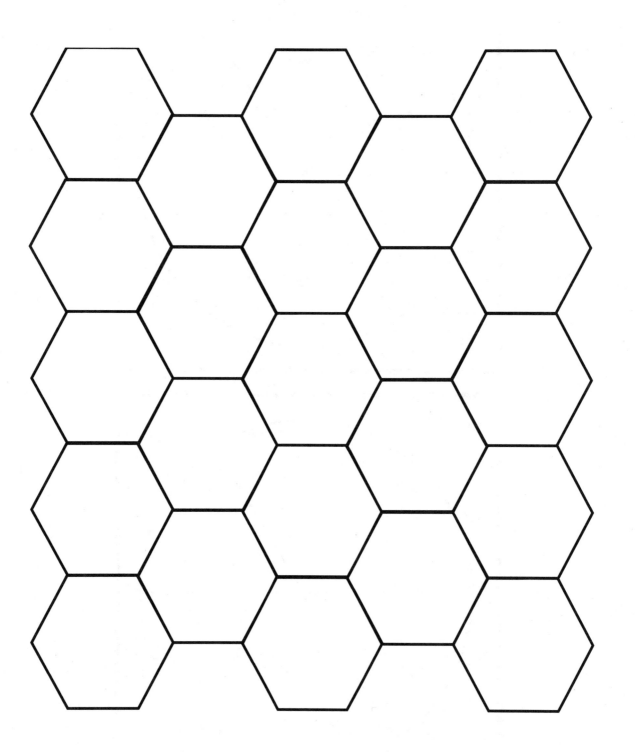

Date _____   Name _____

Subtraction Squares

Subtract each row and column to fill in the missing numbers.

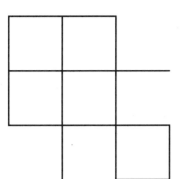 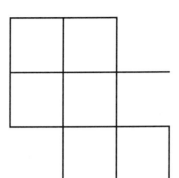 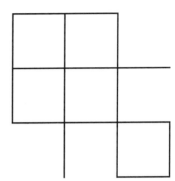

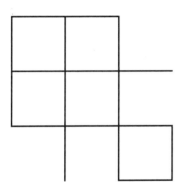 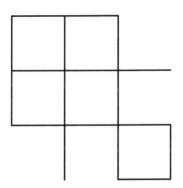 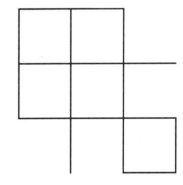

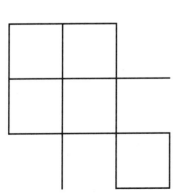 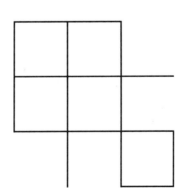 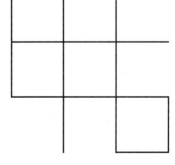

Blackline Masters 155

Ten Frames

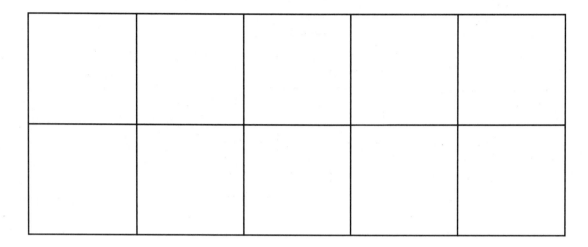

Addition Chart

+	1	2	3	4	5	6	7	8	9	10
1	2	3	4	5	6	7	8	9	10	11
2	3	4	5	6	7	8	9	10	11	12
3	4	5	6	7	8	9	10	11	12	13
4	5	6	7	8	9	10	11	12	13	14
5	6	7	8	9	10	11	12	13	14	15
6	7	8	9	10	11	12	13	14	15	16
7	8	9	10	11	12	13	14	15	16	17
8	9	10	11	12	13	14	15	16	17	18
9	10	11	12	13	14	15	16	17	18	19
10	11	12	13	14	15	16	17	18	19	20

Blackline Masters

Nimble with Numbers Answer Key

p.18 *Just the Facts 1*
1) 7   2) 9   3) 12   4) 16   5) 7   6) 7   7) 11   8) 13   9) 16   10) 14
Go On   16, 20

Just the Facts 2
1) 7   2) 10   3) 12   4) 17   5) 8   6) 9   7) 11   8) 13   9) 16   10) 16
Go On   Samples 5 + 6, 3 + 8, 2 + 9

p. 19 *Just the Facts 3*
1) 8   2) 10   3) 11   4) 15   5) 8   6) 9   7) 10   8) 12   9) 13   10) 18
Go On   20, 25, 30

Just the Facts 4
1) 6   2) 10   3) 12   4) 14   5) 6   6) 9   7) 12   8) 13   9) 15   10) 15
Go On   Samples 6 + 6, 5 + 7, 4 + 8

p. 20 *Just the Facts 5*
1) 4   2) 9   3) 12   4) 14   5) 9   6) 9   7) 11   8) 13   9) 14   10) 17
Go On   12, 15, 18

Just the Facts 6
1) 6   2) 10   3) 11   4) 13   5) 7   6) 10   7) 12   8) 12   9) 17   10) 15
Go On   Samples 6 + 7, 5 + 8, 4 + 9

p. 28 *Seeking Sums Practice I*
Answers may vary. Samples given.

1 + 2,   2 + 4,   4 + 5
2 + 2,   2 + 5,   5 + 4 + 1
1 + 4,   4 + 4,   5 + 4 + 2

1 + 3,   3 + 6
1 + 5,   5 + 5
1 + 6,   5 + 6
3 + 5,   5 + 6 + 1

p. 29 *Seeking Sums Practice II*
Answers will vary. Samples are given.

1 + 2,   1 + 6,   4 + 6
1 + 4,   2 + 6,   6 + 4 + 1
2 + 4,   6 + 2 + 1,   6 + 6

1 + 5,   5 + 5
2 + 5,   5 + 6
2 + 6,   6 + 6
6 + 2 + 1,   6 + 5 + 2

p. 30 *Joining Neighbors I*

p. 31 *Joining Neighbors II*

p. 32   *Joining Neighbors III*

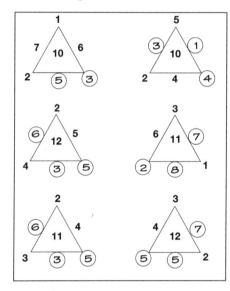

p. 33   *Sum Triangles I*

p. 34   *Sum Triangles II*

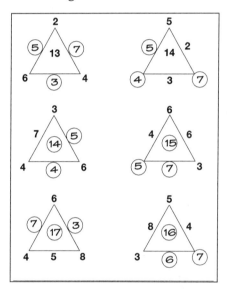

p. 41   **Just the Facts 7**
1) 3   2) 4   3) 8   4) 7   5) 7   6) 7   7) 5   8) 8   9) 7   10) 7
Go On   6

Just the Facts 8
1) 3   2) 4   3) 8   4) 7   5) 4   6) 4   7) 8   8) 6   9) 8   10) 8
Go On   △ = 6,   □ = 4

p. 42   **Just the Facts 9**
1) 3   2) 3   3) 5   4) 6   5) 4   6) 4   7) 2   8) 9   9) 9   10) 9
Go On   10

Just the Facts 10
1) 1   2) 2   3) 7   4) 8   5) 3   6) 3   7) 3   8) 4   9) 9   10) 7
Go On   △ = 8,   □ = 4

p. 43   **Just the Facts 11**
1) 5   2) 8   3) 3   4) 6   5) 3   6) 3   7) 8   8) 6   9) 9   10) 8
Go On   6

p. 43 ctd. *Just the Facts 12*

1) 4   2) 6   3) 5   4) 4   5) 8   6) 8   7) 7   8) 8   9) 7   10) 9

Go On △ = 6, □ = 3

p. 51 *Subtraction Squares I*

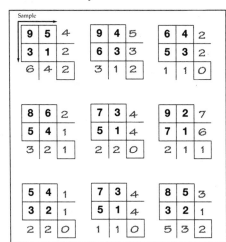

p. 52 *Subtraction Squares II*

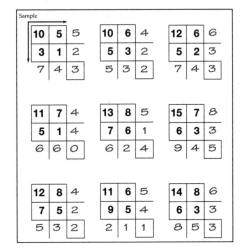

p. 53 *Mystery Numbers Practice I*

6 + 5 = 11,   6 - 5 = 1,   □ = 6,   △ = 5
8 + 4 = 12,   8 - 4 = 4,   □ = 8,   △ = 4
7 + 5 = 12,   7 - 5 = 2,   □ = 7,   △ = 5,
6 + 4 = 10,   6 - 4 = 2,   □ = 6,   △ = 4
8 + 2 = 10,   8 - 2 = 6,   □ = 8,   △ = 2
7 + 4 = 11.   7 - 4 = 3,   □ = 7,   △ = 4

p. 54 *Mystery Numbers Practice II*

11 + 6 = 17,   11 - 6 = 5,   □ = 11,   △ = 6
14 + 4 = 18,   14 - 4 = 10,   □ = 14,   △ = 4
12 + 4 = 16,   12 - 4 = 8,   □ = 12,   △ = 4
13 + 4 = 17,   13 - 4 = 9,   □ = 13,   △ = 4
12 + 5 = 17   12 - 5 = 7,   □ = 12,   △ = 5
13 + 5 = 18   13 - 5 = 8,   □ = 13,   △ = 5

p. 55 *Making Equations Practice I*

1) 4 - 2 - 1 = 1   2) 4 + 1 - 2 = 3   3) 6 - 3 + 1 = 4   4) 6 - 3 - 1 = 2   5) 5 + 3 - 1 = 7
6) 5 + 1 - 3 = 3   7) 4 + 3 - 2 = 5   8) 4 - 3 + 2 = 3   9) 5 + 3 - 1 = 7   10) 5 - 3 + 1 = 3

p. 56 *Making Equations Practice II*

1) 8 - 5 + 4 = 7   2) 5 + 4 - 8 = 1   3) 5 - 4 + 8 = 9   4) 8 + 7 - 9 = 6   5) 9 + 7 - 8 = 8
6) 8 - 7 + 9 = 10   7) 9 - 4 + 5 = 10   8) 9 + 4 - 5 = 8   9) 9 + 4 + 5 = 18,
Equations: 9 + 6 + 2 = 17,   9 - 6 + 2 = 5,   9 + 6 - 2 = 13,   9 - 6 - 2 = 1

p. 62 *Sensible Cents 1*

1) DDP   2) QDP   3) 30¢   4) 41¢   5) DD   6) DNN   7) 6¢   8) 5¢
9–10) Answers will vary.   Go On Answers will vary.

p. 63 *Sensible Cents 2*

1) DDN   2) QQN   3) 21¢   4) 80¢   5) DN   6) QQP   7) 6¢   8) 5¢
9–10) Answers will vary.   Go On   QDD

p. 64 *Sensible Cents 3*

1) QDD   2) QPP   3) 25¢   4) 56¢   5) QP   6) QQD   7) 4¢   8) 7¢
9–10) Answers will vary.   Go On Sample: □ □ □ O △ △ △

p. 70 *Buying Snacks*

1) QDD   2) QQDNN   3) DDDDPP   4) DDNNN   5) QDDNP
6) QDDDN   7) QQQD   8) QDPPP

p. 71   *Find the Combination*
    1) QPP   2) DNP   3) QDP   4) QDN   5) DNPP
    6) DDNP   7) DNNN   8) DNNP   9) DDNN   10) DDDN

p. 72   *What's My Change?*
    1) 25¢; 3¢   2) 30¢; 3¢   3) 45¢; 7¢   4) 50¢; 21¢   5) 35¢; 4¢
    6) 70¢; 8¢   7) 75¢; 20¢

p. 81   *What's in That Place? 1*
    1) 32, 33, 43   2) 21, 31, 32   3) 24, 42, 57   4) 37   5) 56
    6) 25   7) 54   8) 53   9) 34   10) 53 or 54
    Go On   52, 62

What's in That Place? 2
    1) 54, 55, 63, 64   2) 22, 32, 42   3) 46, 50, 63   4) 45   5) 72
    6) 57   7) 42   8) 47   9) 24   10) 74
    Go On   24; It is only one more than 23.

p. 82   *What's in That Place? 3*
    1) 65, 74, 75, 84   2) 55, 65, 66, 67   3) 62, 64, 84   4) 48   5) 63
    6) 42   7) 75   8) 76   9) 57   10) 56
    Go On   44

What's In That Place? 4
    1) 46, 47, 57   2) 18, 28, 29   3) 27, 36, 43   4) 29   5) 54
    6) 51   7) 24   8) 21   9) 42   10) 42
    Go On   Answers will vary.

p. 83   *What's in That Place? 5*
    1) 44, 52, 53, 54   2) 48, 58, 68   3) 18, 25, 52   4) 15   5) 82
    6) 54   7) 89   8) 98   9) 48   10) 84
    Go On   54

What's in That Place? 6
    1) 71, 81, 90, 91   2) 62, 63, 73, 74   3) 39, 43, 52   4) 36   5) 89
    6) 48   7) 67   8) 76   9) 26   10) 27
    Go On   37

p. 90   *What Numbers Are Missing? I*

p. 91   **What Numbers Are Missing? II**   p. 92   **What Numbers Are Missing? III**

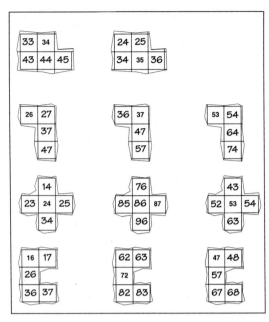
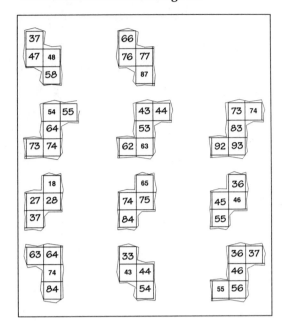

p. 93   *Staircase Number Puzzle I*
   1) 53   2) 36   3) 68   4) 85   5) 59   6) 92   7) 24   8) 46   9) 67

p. 94   *Staircase Number Puzzle II*
   1) 57   2) 74   3) 46   4) 61   5) 18   6) 85   7) 52   8) 27   9) 73   10) 34   11) 49

p. 95   *Creating Numbers I*
   Answers may vary. Samples are given.
   1) 23   2) 23   3) 52   4) 23   5) 43   6) 34   7) 94   8) 82   9) 52   10) 28

p. 96   *Creating Numbers II*
   Answers may vary. Samples are given.
   1) 671   2) 416   3) 147   4) 146   5) 257   6) 725
   7) 875   8) 852 or 872   9) 613   10) 136   11) 361   12) 936

p. 102   *Sum It Up 1*
   1) 21   2) 30   3) 41   4) 53   5) 79   6) 42   7) 56   8) 50   9) 82   10) 142
   Go On   27, 30

   Sum It Up 2
   1) 25   2) 52   3) 72   4) 32   5) 58   6) 51   7) 65   8) 71   9) 73   10) 154
   Go On   33

p. 103   *Sum It Up 3*
   1) 22   2) 41   3) 63   4) 52   5) 79   6) 71   7) 75   8) 72   9) 85   10) 131
   Go On   Answers will vary.

   Sum It Up 4
   1) 21   2) 33   3) 62   4) 61   5) 59   6) 62   7) 88   8) 82   9) 66   10) 151
   Go On   Answers will vary. All sums should equal 50.

p. 104   *Sum It Up 5*
   1) 24   2) 50   3) 34   4) 43   5) 89   6) 53   7) 74   8) 73   9) 87   10) 172
   Go On   Answers will vary.

   Sum It Up 6
   1) 22   2) 43   3) 41   4) 81   5) 77   6) 72   7) 97   8) 70   9) 84   10) 163
   Go On   43, 48

162   *Nimble with Numbers*

p. 112 *Loop Addition I*

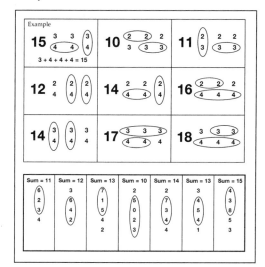

p. 113 *Loop Addition II*

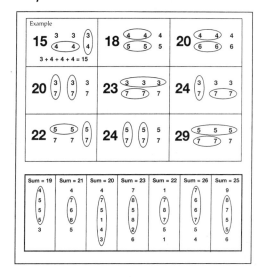

p. 114 *Addition Trees I*

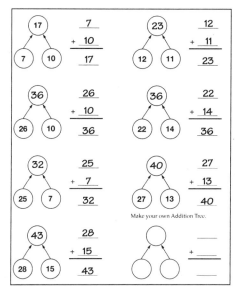

p. 115 *Addition Trees II*

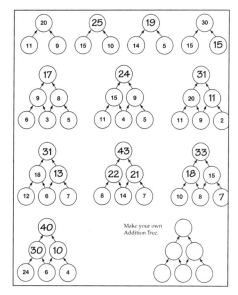

p. 116 *Making Sums I*

Answers may vary.

10 + 17 = 27;	10 + 30 = 40
20 + 30 = 50	10 + 20 = 30
20 + 25 = 45	15 + 25 = 40
20 + 15 = 35	15 + 10 = 25
10 + 35 = 45	20 + 35 = 55
35 + 9 = 44	10 + 9 = 19
15 + 30 = 45	15 + 20 = 35
7 + 15 = 22	20 + 30 = 50

p. 117 *Making Sums II*

Answers may vary.

10 + 20 = 30	17 + 15 = 32
20 + 15 = 35	15 + 10 + 20 = 45
10 + 20 = 30	10 + 15 = 25
15 + 9 = 24	15 + 10 + 9 = 34
30 + 11 = 41	25 + 15 = 40
30 + 15 = 45	11 + 15 = 26
25 + 15 + 11 = 51	25 + 30 + 15 = 70
20 + 25 = 45	30 + 9 = 39
20 + 9 + 30 = 59	30 + 20 + 25 = 75

p. 118 *Searching for 30*

Across	14 + 6 + 10 = 30;	13 + 6 + 11 = 30;	7 + 9 + 14 = 30;	8 + 12 + 10 = 30
Down	13 + 9 + 8 = 30;	10 + 6 + 14 = 30;	11 + 7 + 12 = 30;	15 + 6 + 9 = 30
Diagonals	7 + 13 + 10 = 30;	15 + 9 + 6 = 30;	10 + 11 + 9 = 30	

Answer Key   163

p. 127   **What's the Difference? 1**
1) 18   2) 9   3) 45   4) 7   5) 13   6) 38   7) 29   8) 39   9) 138   10) 214
Go On   68

What's the Difference? 2
1) 25   2) 43   3) 23   4) 6   5) 25   6) 38   7) 21   8) 39   9) 245   10) 127
Go On   97

p. 128   **What's the Difference? 3**
1) 38   2) 38   3) 24   4) 5   5) 4   6) 55   7) 36   8) 16   9) 328   10) 123
Go On   51

What's the Difference? 4
1) 8   2) 34   3) 63   4) 12   5) 24   6) 45   7) 12   8) 28   9) 184   10) 125
Go On   Answers will vary.

p. 129   **What's the Difference? 5**
1) 12   2) 18   3) 29   4) 8   5) 23   6) 45   7) 14   8) 36   9) 239   10) 147
Go On   78

What's the Difference? 6
1) 26   2) 34   3) 19   4) 9   5) 2   6) 37   7) 22   8) 8   9) 444   10) 133
Go On   Answers will vary.

p. 139   **Subtraction Squares III**

Sample
36	20	16
19	7	12
17	13	4

54	30	24
49	25	24
5	5	0

37	20	17
18	12	6
19	8	11

42	39	3
31	28	3
11	11	0

38	29	9
15	12	3
23	17	6

46	30	16
27	15	12
19	15	4

56	21	35
19	7	12
37	14	23

45	19	26
35	15	20
10	4	6

28	9	19
19	6	13
9	3	6

p. 140   **Subtraction Squares IV**

Sample
60	19	41
23	14	9
37	5	32

65	39	26
29	15	14
36	24	12

71	23	48
27	18	9
44	5	39

74	21	53
49	17	32
25	4	21

64	31	33
36	25	11
28	6	22

57	29	28
38	19	19
19	10	9

81	41	40
25	17	8
56	24	32

70	35	35
34	18	16
36	17	19

72	45	27
24	19	5
48	26	22

p. 141   **Finding Pairs I**
Answers may vary. Samples are given.
66 - 56 = 10   66 - 36 = 30
56 - 36 = 20   66 - 26 = 40
30 - 25 = 5   40 - 25 = 15
40 - 30 = 10   40 - 15 = 25
44 - 14 = 30   24 - 9 = 15
14 - 9 = 5   44 - 9 = 35

p. 142   **Finding Pairs II**
Answers may vary. Samples are given.
75 - 25 = 50   25 - 15 = 10
75 - 15 = 60   75 - 50 = 25
51 - 21 = 30   21 - 7 = 14
51 - 7 = 44   31 - 7 = 24
49 - 29 = 20   49 - 11 = 38
39 - 11 = 28   29 - 11 = 18
35 - 16 = 19   65 - 35 = 30
55 - 16 = 39   65 - 16 = 49

p. 143   **Rearrange and Find I**
46 - 2 = 44   26 - 4 = 22   24 - 6 = 18
62 - 4 = 58   64 - 2 = 62   42 - 6 = 36
45 - 3 = 42   53 - 4 = 49   43 - 5 = 38

p. 144   **Rearrange and Find II**
Answers may vary. Samples are given.
23 - 5 = 18   52 - 3 = 49   23 + 5 = 28
53 - 2 = 51   25 - 3 = 22   52 + 3 = 55
32 - 5 = 27   35 + 2 = 37   35 - 2 = 33